AT LAST!! ENCODED TOTALS SECOND ADDITION

AT
LAST!!
ENCODED
TOTALS
SECOND
ADDITION

```
            A T
        L A S T !!
  E N C O D E D
    T O T A L S
    S E C O N D
---------------
A D D I T I O N
```

The long-awaited sequel to
HAVE SOME SUMS TO SOLVE

STEVEN KAHAN

Library of Congress Catalog Card Number 94-16801.
ISBN: 0-89503-171-X (paper)

Library of Congress Cataloging-in-Publication Data

Kahan, Steven.
At last!! : encoded totals second addition / by Steven Kahan.
p. cm.
1. Mathematical recreations. I. Title.
QA95.K32 1994
793.7'4- -dc20 94-16801
CIP

dedication
directed approach: page 85
solution: page 119

A	C	E	F	H	I	R	S	T	Y
0	1	2	3	4	5	6	7	8	9

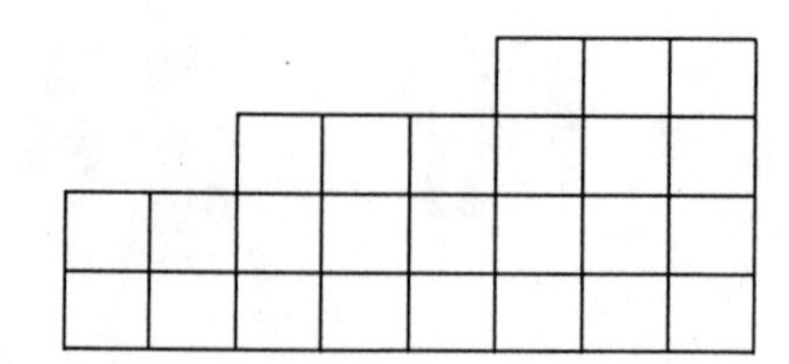

dedication

For my wife, Susan, and for my children, David and Sara. This incredible trio has taught me the true meanings of four common words – friendship, trust, respect, and love.

Y E S
T H E Y'R E
T E R R I F I C
T E A C H E R S

acknowledgment

Some of the ideal doubly-true alphametics have previously appeared in print on the pages of the *Journal of Recreational Mathematics*, a quality publication issued quarterly by

Baywood Publishing Company, Inc.
26 Austin Avenue
P. O. Box 337
Amityville, New York 11701

Sincere gratitude is hereby expressed to Stuart Cohen, its president, who graciously allowed me to reprint these, as well as some expository material from the preface of my first collection of alphametics, HAVE SOME SUMS TO SOLVE. Information concerning both *JRM* and *HSSTS* can be obtained by writing to Baywood at the above address.

foreword

Alphametics or cryptarithms—mathematicians have yet to settle on the favored term—is a branch of recreational number theory that is closely related to cryptograms. In solving a cryptogram, one combines a knowledge of language with shrewd reasoning; in solving an alphametic, one combines a knowledge of elementary number theory with shrewd reasoning. In both cases, there is great intellectual satisfaction that comes with "cracking the code" and finding the solution.

Curiously, the technique of solving an alphametic models in many ways the process by which laws of science (nature's puzzles) are discovered. When testing a conjecture, a scientist searches for supporting physical evidence. As such evidence accumulates, a point is reached at which the scientist suddenly becomes certain that his conjecture will be confirmed. There are similar turning points in solving an alphametic. At some juncture in the testing process, everything starts to drop neatly into place, and one has the pleasure of knowing that a solution is near at hand. In science, of course, the solution is never absolutely certain, and it is here that the analogy breaks down. When you solve an alphametic, you *know* that the solution is the correct one.

Computer programs are available for solving alphametics, but surely they steal the exhilaration from the whole process. What satisfaction can one get from solving, say, a chess problem by giving it to a computer that will crack it in a matter of seconds? Using a desk calculator to cut the time spent in solving an alphametic is quite a different thing, but a computer program seems to sabotage the mission. The whole enjoyment in solving an alphametic is to do so by hand.

Although mathematics journals and puzzle magazines often include alphametics, not many books have been devoted entirely to the topic. I know of only three in English: Maxey Brooke's *One Hundred and Fifty Problems in Crypt-Arithmetic* (1963); *Cryptarithms* (1976), by Josephine and Richard Andre; and Steven Kahan's previous book, *Have Some Sums To Solve* (1978). For fifteen years, Kahan has edited the alphametics section of the *Journal of Recreational Mathematics*. This new book, like his earlier one, includes a few choice problems selected from that admirable quarterly, but for the most part presents previously unpublished material. To make the book even more interesting, Kahan has interspersed his alphametics with short paragraphs about fascinating facts in number theory.

Another aspect of alphametic play is the actual construction of good problems. Optimally, one attempts to use each available digit, to code the numbers with words that make sense, and to create a puzzle that can be uniquely solved by logical reasoning as opposed to mere trial-and-error. This, of course, is not always possible—you have only to try making one to discover how difficult the task can be. Few can equal Kahan in the ability to devise such elegant alphametics.

At Last!! Encoded Totals Second Addition is a superb collection sure to delight every alphametic buff. I suspect it will also introduce many a reader to a flourishing subset of recreational mathematics that he or she may not have known about before.

Martin Gardner

table of contents

interesting integer idiosyncracies can be found on the following pages:

9, 15, 16, 28, 30, 33, 35, 39, 47, 54, 57,
59, 63, 65, 67, 72, 75, 77, 79.

preface

Creating alphametics is very much like eating good grapes—once you find a sweet bunch, you simply cannot resist going back to seek another, as good if not better than the first. Since the publication of my first collection of alphametics, *Have Some Sums To Solve*, back in 1978, various new ideas have arisen from time to time. These have led to nicely phrased puzzles that are solvable in a reasonable way—the prerequisites for what I consider to be a "good" alphametic. As is my wont, I scrupulously jot down such items, file the paper away, and go on with the less enjoyable, more mundane everyday tasks that are required of us all. One such chore, a purge of the material that collects in and on my desk, is periodically performed once every decade (or once every score, if I tend to procrastinate a bit!). On doing so recently, I came across a folder ominously labeled "For The Future." Recalling the credo that the future is now, I cautiously opened the folder's cover. Cascading out came a mass of loose sheets, which after considerable spit and polish, has turned into the volume that you now hold in your hands.

If you are already an experienced alphametician, feel free to skip the remainder of this preface—you are no doubt aware of its contents. For the tyro, however, some further exposition might prove helpful, and it is provided with that hope in mind.

An alphametic is a thought-provoking decoding puzzle that requires the solver to replace each letter in the puzzle with an appropriately chosen digit so that the decoded version is a valid arithmetic statement. Essentially, there are three rules that govern the solution process:

1. once a digit is assigned to a letter, that digit must be assigned to every appearance of that letter in the puzzle;
2. once a digit is assigned to a letter, that digit cannot be assigned to any other letter in the puzzle; and
3. the digit zero cannot be assigned to a letter that appears at the beginning of any word in the puzzle.

The forty puzzles presented in Section 1, as well as the cover and dedication puzzles, all fall into the special subcategory of additive alphametics. Each of their sums has a unique decoding, sometimes insured by the imposition of an initial condition. The presence of a pair of interchangeable digits within the summands themselves does not constitute a contradiction of this uniqueness.

Within the subcategory, two varieties of alphametics are included—the ideal, doubly-true type (abbreviated "i.d.t.") and the narrative type. In the former, all ten digits appear in the solution, and a mathematically correct addition example results when the problem is read aloud. The latter type is presented within the context of a brain-teaser (which itself requires solution) and/or some informative paragraphs. Throughout this section will also be found some "integer idiosyncracies" to tantalize the reader's mathematical taste buds.

No explicit method can be outlined for a general solution procedure, since each alphametic necessitates its own individual approach. Be forewarned, though, that an inappropriate entry into a problem can lead to a hopeless morass of trial-and-error calculation. Much of this can be bypassed through a combination of foresight, perception, and finesse. Here are some broad hints that are often useful:

1. If a summand or sum is divisible by an even number, then the letter in its units column cannot represent an odd integer.
2. If a summand or sum is a prime, then the letter in its units column must represent either 1, 3, 7, or 9.
3. If a summand or sum is a perfect square, then the letter in its units column cannot represent 2, 3, 7, or 8.
4. If a summand or sum is divisible by 5, then the letter in its units column must represent either 0 or 5.
5. If a summand or sum is divisible by 10, then the letter in its units column must represent 0.
6. Always be sure to consider the magnitude of any potential carryover from one column into the column to its immediate left. Whenever possible, establish constraints on the magnitude of a digit represented by a particular letter, based upon the location of that letter in the puzzle.
7. If possible, determine the parity of the digit represented by a particular letter. For instance, suppose that an alphametic contains precisely two summands. If the same letter appears in the tens column of each summand and if there is a known carryover of 1 from the units column, then the letter in the tens column of the sum must represent an odd digit.
8. If a particular column contains five copies of the same letter, then that portion of the column must add up to an integer which ends in either 0 or 5. As an example, suppose that an alphametic contains precisely five summands. If the same letter appears in the hundreds column of each summand and if there is a known carryover of 2 from the tens column, then the letter in the hundreds column of the sum must represent either 2 or 7.
9. If the same letter appears in the column of a summand and also in the corresponding column of the sum, then the remaining entries in that column plus any carryover from the previous column must add up to an integer which ends in 0. To illustrate, suppose that an alphametic contains precisely three summands, each of which has a different letter representing

its units digit. Assume further that the letter representing the units digit of the sum is identical to that of one of the summands. Then the remaining two entries in the units column must add up to 10, which immediately yields a carryover of 1 into the tens column.

10. If a particular column contains ten copies of the same letter, simply erase them all and put a single copy of that letter into the column immediately to the left of the original. This maneuver is called a ten-shift. In a similar fashion, if a particular column contains one hundred copies of the same letter, perform a hundred-shift by removing them all and placing a single copy of that letter into the column that is two to the left of the original.

Section 2 offers directed approaches to each of the puzzles. These discussions are tailored to provide some strategic guidance without removing the challenge associated with the quest for the actual answer. Solutions to all puzzles are presented in Section 3, these being randomly ordered to avoid inadvertent views of the "next result." This section also contains responses to all questions raised within the context of the narrative alphametics. Lastly, a solutions chart is given in order to inform the interested reader how many ways exist to solve each puzzle if no initial condition were imposed.

The appeal of these puzzles can really be traced to the fact that achieving success is virtually independent of one's mathematical prowess. Logical thought, cleverness, and tenacity are the major weapons used to unravel an alphametic. Surely you have those in your arsenal, don't you? Then get to work and have fun!

? ? ? ? ? ? ? ? ? ? ? ? ? ? ? ? ? ? ? ?

SECTION 1

PUZZLES

? ? ? ? ? ? ? ? ? ? ? ? ? ? ? ? ? ? ? ?

even odds
directed approach: page 86
solution: page 115

A	C	E	F	H	I	N	T	Y
1	2	3	4	5	6	7	8	9

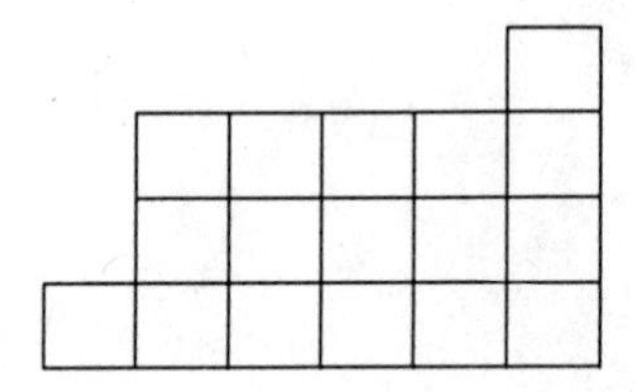

even odds

Two words that have different sounds and meanings but have identical spellings are called heteronyms. If you ran across a word such as "sewer" out of context and were asked to pronounce it, you would have only

```
          A
  F I F T Y
  F I F T Y
-----------
C H A N C E
```

of doing so correctly.

Decode this alphametic without using the digit zero in your solution. Then consider doing something more.

THE OBJECT OF THIS QUIZ IS TO RECORD ALL OF THE HETERONYMS THAT APPEAR IN THIS PARAGRAPH. WHERE DOES THIS SEARCH LEAD? WELL, IF YOU READ VERY, VERY CAREFULLY, AND YOU USE ALL OF YOUR POWERS OF OBSERVATION, YOU SHOULD HAVE NO EXCUSE BUT TO FIND THE TOTAL NUMBER OF HETERONYMS PRESENT HERE. IF YOU WIND UP GETTING THEM ALL, TAKE A BOW; IF NOT, DON'T SHED A TEAR!

How many have you discovered?

go for the gold
directed approach: page 86
solution: page 117

C	E	G	H	I	L	N	S	T	W
0	1	2	3	4	5	6	7	8	9

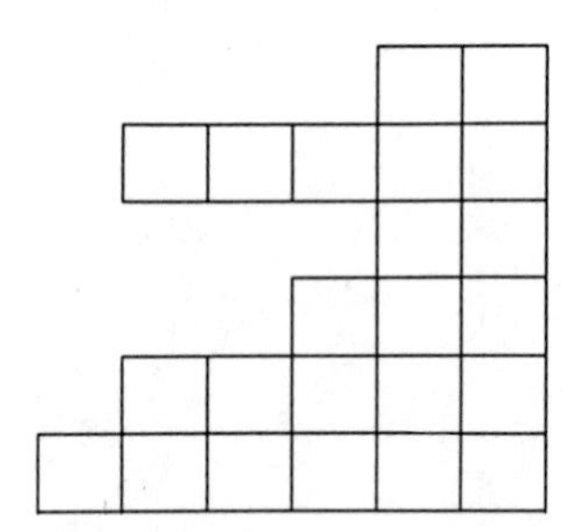

go for the gold

Here's a chance to be rewarded for cleverness. You are confronted with nine identical-looking bags, each of which contains ten gold nuggets. If the truth be told, however, only one of these bags contains real gold; the other eight are filled with worthless fool's gold. A nugget of fool's gold is known to weigh exactly 1 ounce, whereas a nugget of real gold weighs precisely $\frac{9}{10}$ of an ounce. Available to you is an ordinary single-pan scale, with which you are permitted to perform one and only one weighing. With nine bags before you, you can reap quite a windfall if you can determine

```
      I N
  W H I C H
        I S
      T H E
  L I G H T
-----------
W E I G H T.
```

The product of four consecutive positive integers can never be equal to a perfect square.

---

i.d.t. – ninety

directed approach: page 87
solution: page 119

E	G	H	I	N	R	S	T	X	Y
0	1	2	3	4	5	6	7	8	9

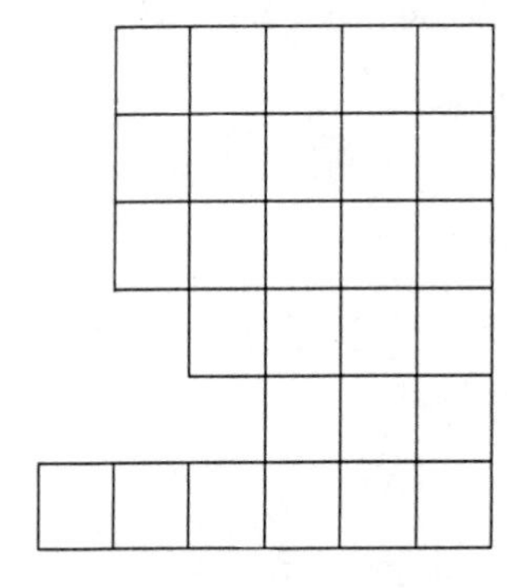

i.d.t. – twenty

directed approach: page 87
solution: page 116

E	H	L	N	O	R	T	V	W	Y
0	1	2	3	4	5	6	7	8	9

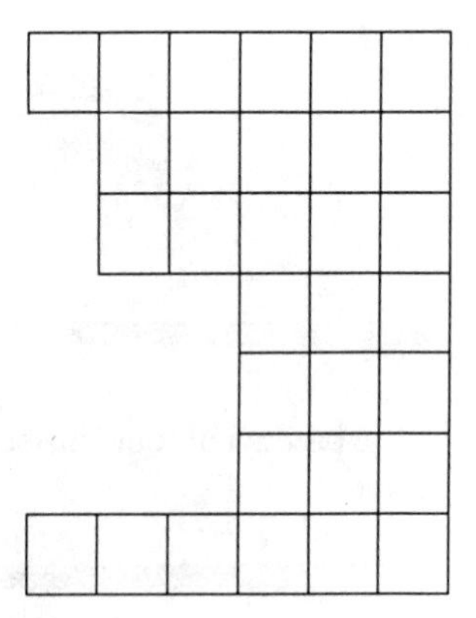

i.d.t. – ninety

```
  S I X T Y
  E I G H T
  T H R E E
    N I N E
      T E N
-----------
N I N E T Y
```

i.d.t. – twenty

```
E L E V E N
  T H R E E
  T H R E E
      O N E
      O N E
      O N E
-----------
T W E N T Y
```

wholes with holes
directed approach: page 88
solution: page 114

F M O R S U Y

0 1 2 3 4 5 6 7 8 9

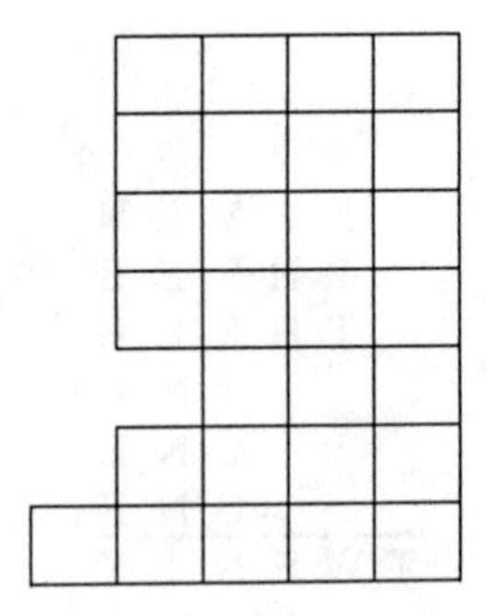

wholes with holes

A perplexing geometric construction problem involves the positioning of four congruent quadrilaterals to create various configurations. Trace the shape shown below and cut out four copies of it.

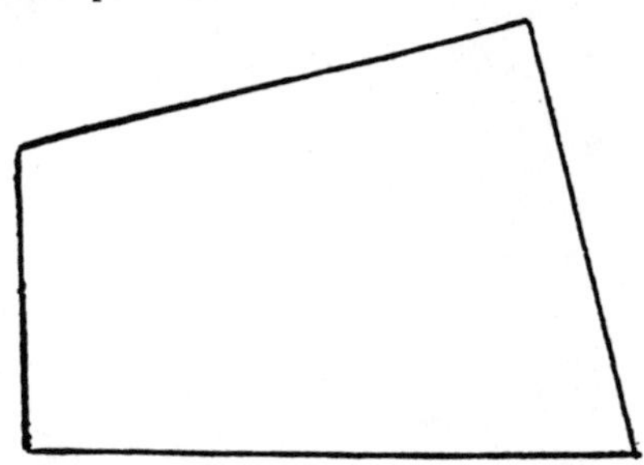

Then arrange these four pieces into the following four patterns:

a. a square with a small square cut from its center;
b. a square with a large square cut from its center;
c. a rectangle with a rectangle cut from its center; and
d. a parallelogram with a parallelogram cut from its center.

If you accept this challenge, then

F R O M
Y O U R
F O U R,
F O R M
O U R
F O U R
F O R M S.

i.d.t. – sixtyone
directed approach: page 88
solution: page 120

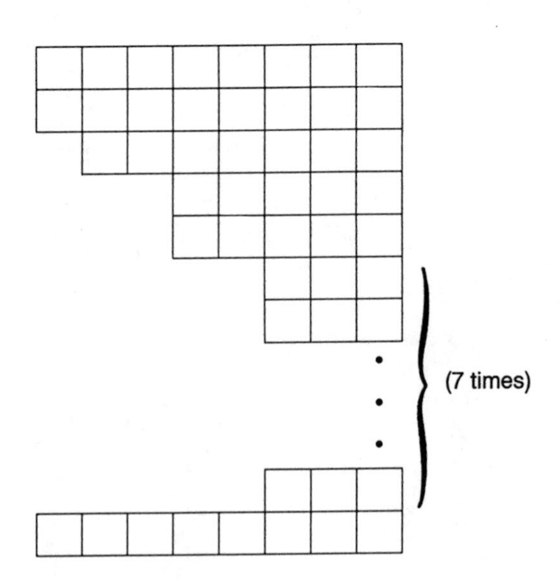

i.d.t. – sixtyone

```
N I N E T E E N
T H I R T E E N
    S I X T E E N
        T H R E E
        T H R E E
            O N E
            O N E
            O N E
            O N E
            O N E
            O N E
            O N E
_________________
S I X T Y O N E
```

$$1^2 + 2^2 + 3^2 + \ldots + 24^2 = 4{,}900 = 70^2$$

For no other value of n is it true that the sum of the first n perfect squares is itself a perfect square.

1,634 is the smallest four-digit number that equals the sum of the fourth powers of its digits.

54,748 is the smallest five-digit number that equals the sum of the fifth powers of its digits.

548,834 is the smallest six-digit number that equals the sum of the sixth powers of its digits.

1,741,725 is the smallest seven-digit number that equals the sum of the seventh powers of its digits.

24,678,050 is the smallest eight-digit number that equals the sum of the eighth powers of its digits.

division decisions

The number 27,720 has the distinction of being the smallest positive integer that is divisible by each of the first twelve counting numbers. (A considerably larger number with this property is 12! = 479,001,600.) Based upon the notion of congruence, tests are available that enable us to predict divisibility without actually performing the division. Let us introduce some notation.

We will write N = "$a_na_{n-1}a_{n-2} \ldots a_2a_1a_0$" to represent the integer under consideration, where a_0 is its units digit, a_1 its tens digit, and so forth. Then N is divisible by

2 if and only if a_0 is even;
3 if and only if $a_n + a_{n-1} + a_{n-2} + \ldots + a_2 + a_1 + a_0$ is divisible by 3;
4 if and only if " a_1a_0 " is divisible by 4;
5 if and only if $a_0 = 0$ or 5;
6 if and only if it is divisible by both 2 and 3;
8 if and only if " $a_2a_1a_0$ " is divisible by 8;
9 if and only if $a_n + a_{n-1} + a_{n-2} + \ldots + a_2 + a_1 + a_0$ is divisible by 9;
10 if and only if $a_0 = 0$;
11 if and only if $a_n - a_{n-1} + a_{n-2} - \ldots + (-1)^n a_0$ is divisible by 11; and
12 if and only if it is divisible by both 3 and 4.

The divisibility test for 7 is not very well known. Specifically, N is divisible by 7 if and only if " $a_2a_1a_0$ " – " $a_5a_4a_3$ " + " $a_8a_7a_6$ " – . . . is divisible by 7, where one or two leading zeros need to be introduced if the number of digits of N is not a multiple of 3. For instance, 3,511,539,579,751 must be divisible by 7, since 751 – 579 + 539 – 511 + 003 = 203 is divisible by 7.

division decisions
directed approach: page 89
solution: page 115

B C D E F I O R V W

0 1 2 3 4 5 6 7 8 9

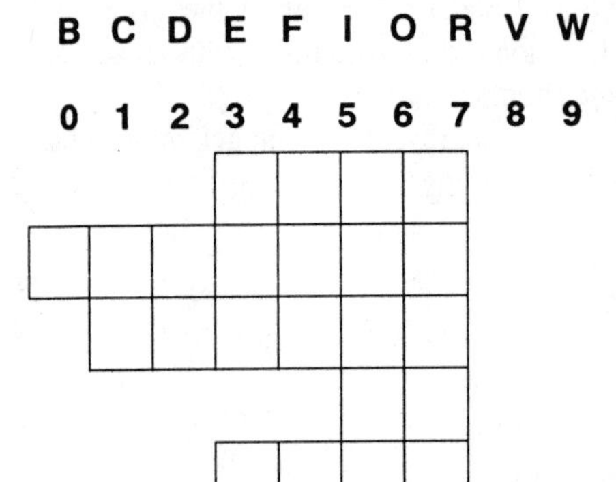

i.d.t. – fiftyone
directed approach: page 89
solution: page 113

E F G H I N O R T Y

0 1 2 3 4 5 6 7 8 9

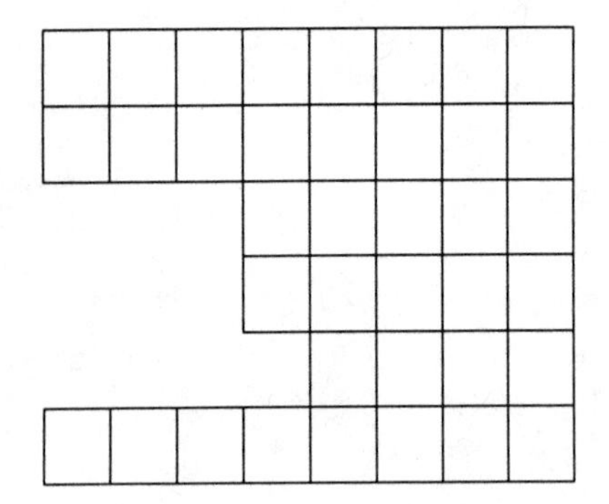

The nicest part about this statement is that

```
          W E' V E
    D E C I D E D
      B E F O R E
              W E
          E V E R
    -------------
    D I V I D E D.
```

Appropriately, use the last stated test to determine the solution in which DIVIDED is divisible by 7.

i.d.t. – fiftyone

```
E I G H T E E N
E I G H T E E N
      T H R E E
      T H R E E
        N I N E
---------------
F I F T Y O N E
```

where THIRTY is even.

optical allusion
directed approach: page 90
solution: page 116

E	I	N	O	S	T	V	W	Y
0	1	2	3	4	5	6	7	8	9

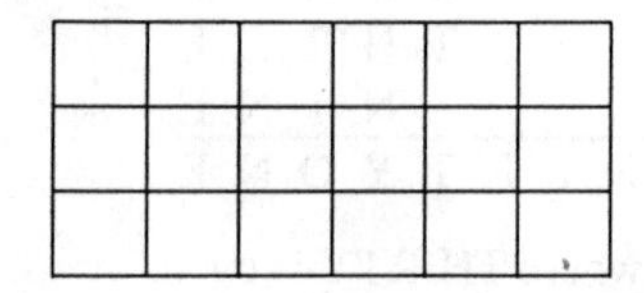

optical allusion

Appearances can certainly be deceiving, as this reprint from a pre-1900 advertisement shows.

If your eyesight and insight are strong, you can literally turn this little frog into another of nature's creatures. It's not a prerequisite, but it certainly wouldn't hurt if you happened to have

$$\frac{\text{T W E N T Y}}{\text{T W E N T Y}}$$
V I S I O N.

open and shut case
directed approach: page 90
solution: page 119

C	D	E	L	N	O	P	R	S	
0	1	2	3	4	5	6	7	8	9

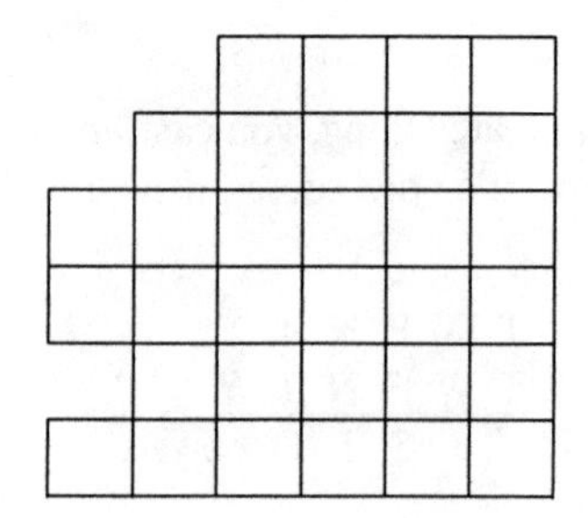

open and shut case

An innovative high school mathematics teacher posed the following problem to his honors class:

"There are one thousand lockers in our school, numbered consecutively from 1 to 1000. At the beginning of the school day, all of the locker doors are closed. Students enter the school in single file. The first student to enter will open each and every locker door. The second student to enter will then close all of the doors on the even-numbered lockers. The third student will next alter the status of the doors on all lockers with numbers that are multiples of 3, followed by the fourth student, who will in turn alter the status of the doors on all lockers numbered with a multiple of 4. This procession will continue until the thousandth and last student enters and completes his or her task. At this time, exactly how many locker doors will be open, and which lockers sport the open doors?"

When asked for a clarification of "altered status," the teacher responded alphametically:

```
    O P E N
  D O O R S
C L O S E D
C L O S E D
  D O O R S
-----------
O P E N E D,
```

where the most OPENED doors are required at the end.

i.d.t. – thousand

directed approach: page 91
solution: page 117

A	D	E	H	N	O	R	S	T	U
0	1	2	3	4	5	6	7	8	9

5 (□□□□□□□) + 10 (□□□) + 400 (□□□) =

□□□□□□□□

i.d.t. – thousand

```
  H U N D R E D
  H U N D R E D
  H U N D R E D
  H U N D R E D
  H U N D R E D
          T E N \
          T E N |
              . |
              . } (10 times)
              . |
          T E N /
          O N E \
          O N E |
              . |
              . } (400 times)
              . |
          O N E /
---------------
T H O U S A N D
```

urban affairs
directed approach: page 92
solution: page 120

A E H I L R S T Y

0 1 2 3 4 5 6 7 8 9

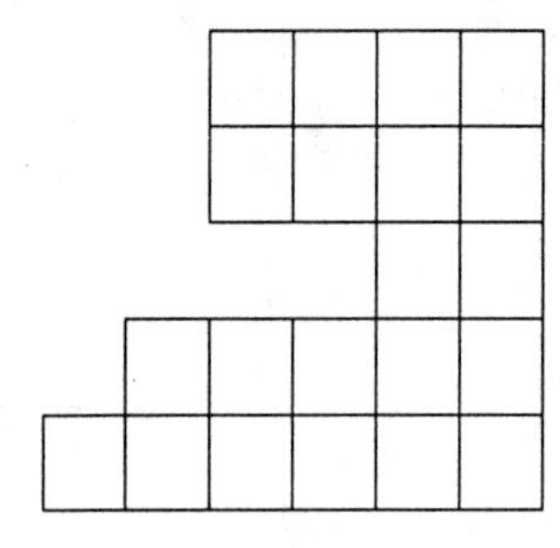

i.d.t. – thirty
directed approach: page 92
solution: page 115

E F H I N R S T V Y

0 1 2 3 4 5 6 7 8 9

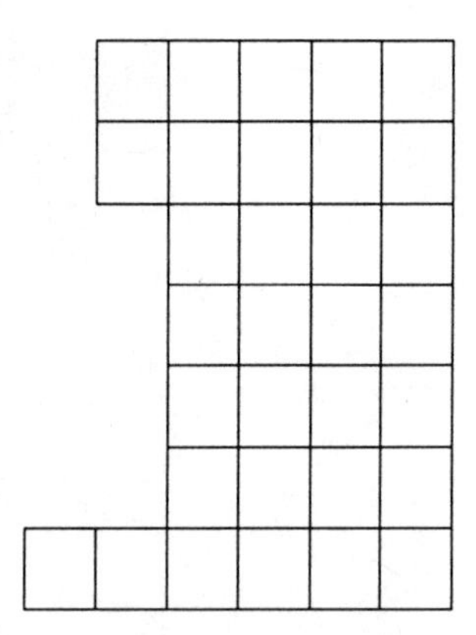

urban affairs

Pick up any newspaper and you are certain to find at least one story headlining the plight of today's big-city dwellers. To be sure, these are serious and complex issues. The problem posed here is in a somewhat lighter vein and is linguistic in nature.

Search through your vocabulary and find as many words as you can that end in -city (other than city itself!). To make matters interesting,

```
   L E T' S
   L I S T
       A T
 L E A S T
-----------
T H I R T Y.
```

i.d.t. – thirty

```
  S E V E N
  T H R E E
    F I V E
    F I V E
    F I V E
    F I V E
-----------
T H I R T Y
```

Given any four consecutive integers greater than 11, there is always at least one which must be divisible by a prime greater than 11.

If digits are removed one by one from the right side of the prime 739,391,133, each of the eight resulting integers will be a prime.

splitting hares

In the early part of the thirteenth century, Leonardo of Pisa, known to his friends as Fibonacci, published a significant text that served as the basis for much of the mathematics that would be developed in the centuries to come. Among its many examples was this one:

"When placed inside a closed room, a pair of newborn rabbits, one male and one female, will bear another pair of rabbits (again one of each gender) in exactly two months. Moreover, the pair will continue to give birth to one such pair of rabbits every month thereafter. If all pairs of rabbits reproduce monogamously according to this scheme, how many pairs will be inside the room one year later?"

To be definite, let's start with the process on January 1. Then the following chart shows the population inside the room on a monthly basis for one year. We refer to a one-month-old pair of rabbits as adolescents and a pair two-months-old or more as adults.

1/1:	1 pair (newborn)
2/1:	1 pair (adolescent)
3/1:	2 pairs (1 adult, 1 newborn)
4/1:	3 pairs (1 adult, 1 adolescent, 1 newborn)
5/1:	5 pairs (2 adult, 1 adolescent, 2 newborn)
6/1:	8 pairs (3 adult, 2 adolescent, 3 newborn)
7/1:	13 pairs (5 adult, 3 adolescent, 5 newborn)
8/1:	21 pairs (8 adult, 5 adolescent, 8 newborn)
9/1:	34 pairs (13 adult, 8 adolescent, 13 newborn)
10/1:	55 pairs (21 adult, 13 adolescent, 21 newborn)
11/1:	89 pairs (34 adult, 21 adolescent, 34 newborn)
12/1:	144 pairs (55 adult, 34 adolescent, 55 newborn)
1/1:	233 pairs (89 adult, 55 adolescent, 89 newborn)

A positive integer can be written as the sum of consecutive positive integers if and only if it is not a power of 2.

There are six two-digit numbers that can be formed from the set {1,2,3} if repetition is not allowed. The sum of these six numbers is 132, one of the permutations of the elements in the given set.

Consequently, the room will be filled with 233 pairs of rabbits after one year has passed.

Being a mathematician and not a farmer, Fibonacci's interest was in the numerical pattern that the solution suggested. Realizing, too, that the problem need not terminate after a year (although rabbits, like people, actually do get old and tired after a while!), he was led to examine the following infinite sequence:

$$1, 1, 2, 3, 5, 8, 13, 21, 34, 55, 89, 144, 233, 377, 610, \ldots$$

This has come to be called the Fibonacci sequence, and each of its terms is called a Fibonacci number. There are many, many interesting and important relationships involving Fibonacci numbers, ten of which are given below. Throughout, we will use the notation F_n to stand for the nth Fibonacci number (so that $F_3 = 2$ and $F_{10} = 55$).

i. For $n \geq 3$, $F_n = F_{n-2} + F_{n-1}$. (This formal definition of the Fibonacci sequence illustrates a recursive relationship. With $F_1 = F_2 = 1$, it simply observes that from the third term on, we can calculate each term in this sequence by adding the previous two.)

ii. $F_1 + F_2 + F_3 + \ldots + F_n + 1 = F_{n+2}$.

iii. $F_1 + F_3 + F_5 + \ldots + F_{2n-1} = F_{2n}$.

iv. $F_2 + F_4 + F_6 + \ldots + F_{2n} + 1 = F_{2n+1}$.

v. $(F_1)^2 + (F_2)^2 + (F_3)^2 + \ldots + (F_n)^2 = F_nF_{n+1}$.

vi. $(F_{n+1})^2 - (F_{n-1})^2 = F_{2n}$.

vii. $F_1F_2 + F_2F_3 + F_3F_4 + \ldots + F_{2n-1}F_{2n} = (F_{2n})^2$.

viii. F_n and F_{n+1} have no common divisor other than 1. (Numbers with this property are called relatively prime.)

ix. If $n > 4$ is not prime, then F_n is not prime.

x. n is a multiple of 5 if and only if F_n is a multiple of 5.

Not everything is known concerning Fibonacci numbers – not by a long shot! For instance, no one yet knows whether there are infinitely many Fibonacci numbers that are prime. This and other open questions are still areas of investigation for today's number theorists.

splitting hares

directed approach: page 93
solution: page 113

A	B	F	H	I	O	P	R	S	T
0	1	2	3	4	5	6	7	8	9

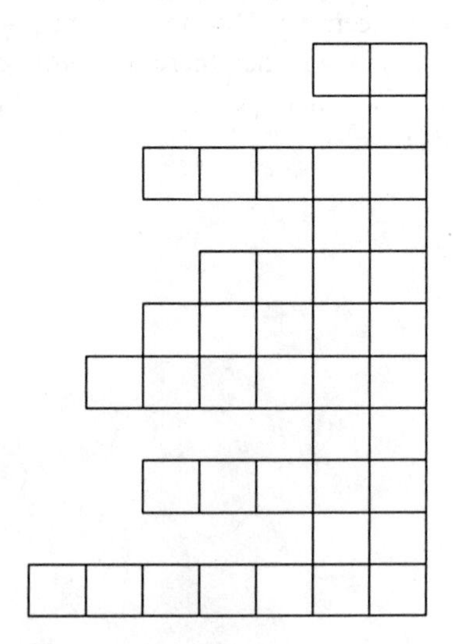

It probably did not take long for Fibonacci to recognize that his original problem was abundantly rich in mathematical treasures. No doubt he was so elated with this observation that he lifted a glass of his finest wine and uttered these words:

```
          S O
            I
    T O A S T
          T O
      F A S T
    B I R T H
  H A B I T S
          O F
    P A I R S
          O F
-------------
R A B B I T S.
```

The integer 6,661,661,161, the square of 81,619, is the largest known perfect square that contains exactly two distinct nonzero digits.

i.d.t. – fortytwo
directed approach: page 94
solution: page 118

E	F	H	N	O	R	T	U	W	Y
0	1	2	3	4	5	6	7	8	9

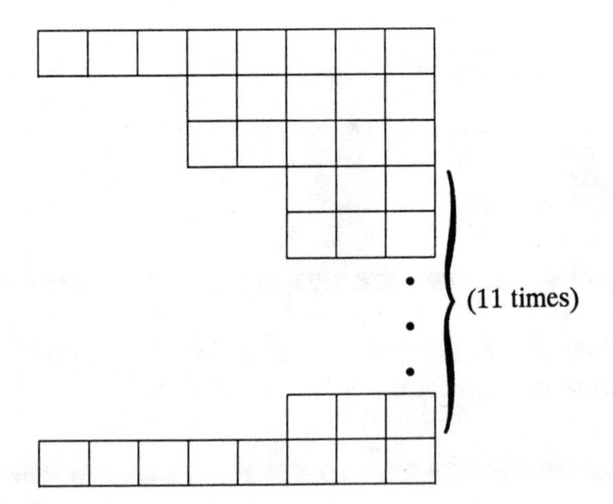

i.d.t. – fortytwo

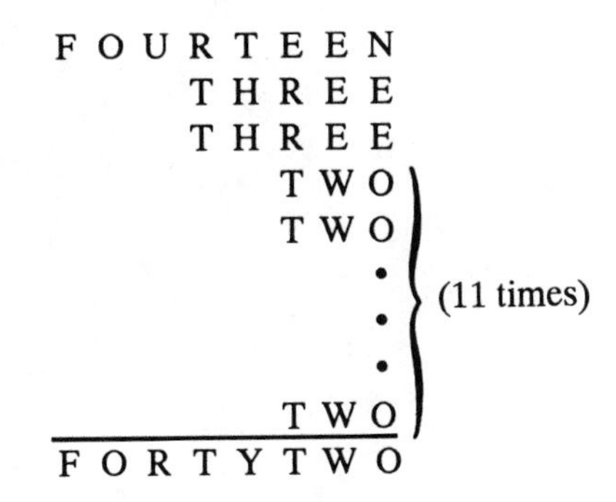

where FORTYTWO is even.

A prime is called absolute if and only if every permutation of its digits is also a prime. The only absolute primes with two or more distinct digits are (13/31), (17/71), (37/73), (79/97), (113/131/311), (199/919/991), and (337/373/733).

---

for the cruciverbalists
directed approach: page 94
solution: page 119

C	D	E	F	N	O	P	R	S	W
0	1	2	3	4	5	6	7	8	9

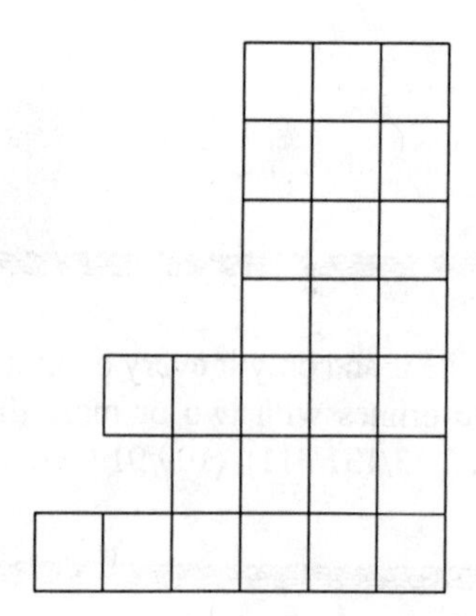

for the cruciverbalists

1	2	3	4
2			
3			
4			

ACROSS
1. take by force
2. employ
3. residents of hives
4. effortlessness

DOWN
1. one is one
2. one thousand is one
3. one million is one
4. one billion is one

Only sixteen letters are required to complete this crossword puzzle, but heed the admonition to proceed with caution! As a hint, you might ponder this friendly warning: if you are wise, you won't let your eyes deceive you, for this puzzle is a bit of a tease. Indeed, it's specifically designed

F O R
O N E
O D D
O D D
C R O S S
W O R D
P E R S O N.

i.d.t. – sixtysix
directed approach: page 95
solution: page 117

E	H	I	N	O	R	S	T	X	Y
0	1	2	3	4	5	6	7	8	9

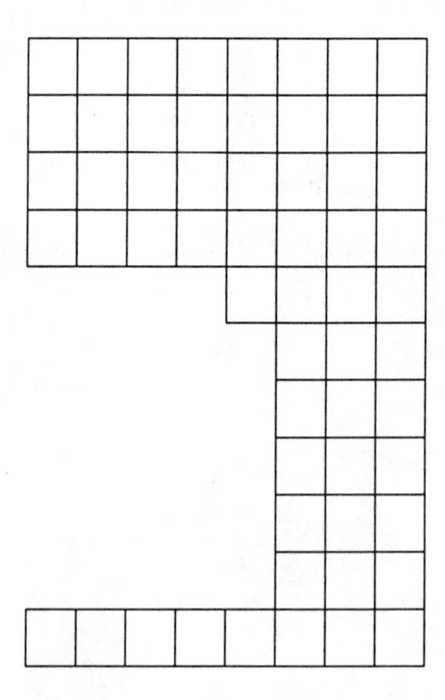

i.d.t. – sixtysix

```
T H I R T E E N
T H I R T E E N
T H I R T E E N
T H I R T E E N
          N I N E
            O N E
            O N E
            O N E
            O N E
            O N E
_______________
S I X T Y S I X
```

1,729 is the smallest positive integer that can be expressed as the sum of two cubes in two different ways:

$$12^3 + 1^3 \text{ and } 10^3 + 9^3.$$

87,539,319 is the smallest positive integer that can be so expressed in three different ways:

$$436^3 + 167^3 = 423^3 + 228^3 = 414^3 + 255^3.$$

count on it!
directed approach: page 95
solution: page 120

E	I	L	N	O	Q	R	S	T	U
0	1	2	3	4	5	6	7	8	9

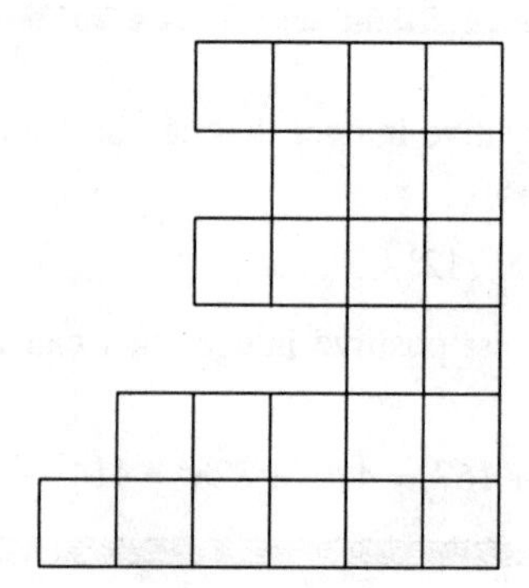

count on it!

The sequence 2 1 2 0 0 is an especially descriptive one. Its first entry, 2, tells us how many times zero appears in the list; its second entry, 1, tells us how many times one appears; its third entry, 2, tells us how many times two appears. Finally, its last two entries indicate that no threes or fours appear in the list at all. As it turns out, this is the only such prophetic sequence of length five that can be constructed.

Now it's your turn! You are hereby requested to create a similarly prophetic sequence that is twice the length of the model. In other words, you must find a set of nonnegative integers so that the first counts the number of zeros in your list, the second counts the number of ones, and so forth. To complete the process, the tenth and final integer should count the number of nines in your list. As is the case with the model, this construction also can be accomplished in exactly one way. Indeed, it is

```
  T R U E –
    O U R
  L I S T
      I S
Q U I T E
-----------
U N I Q U E.
```

i.d.t. – fiftytwo
directed approach: page 96
solution: page 113

E	F	I	N	O	S	T	W	X	Y
0	1	2	3	4	5	6	7	8	9

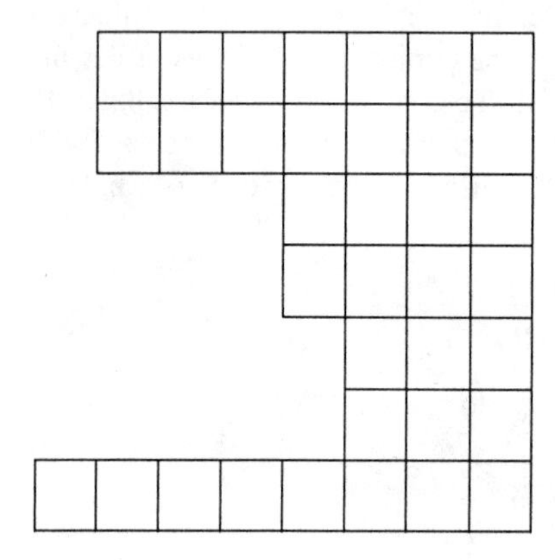

i.d.t. – seventy
directed approach: page 96
solution: page 118

E	H	L	N	R	S	T	V	W	Y
0	1	2	3	4	5	6	7	8	9

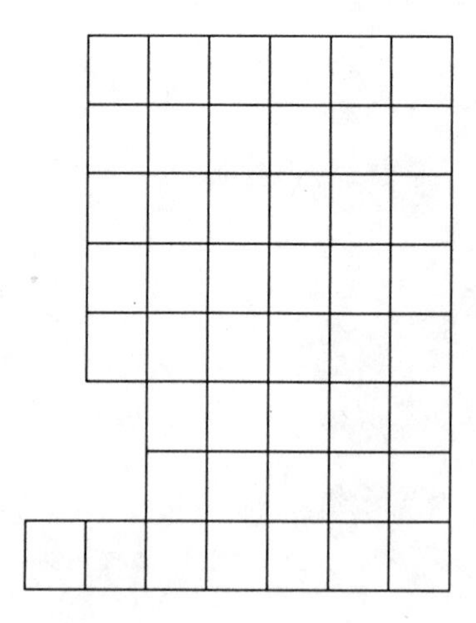

i.d.t. – fiftytwo

```
  S I X T E E N
  S I X T E E N
        N I N E
        N I N E
          O N E
          O N E
  -------------
F I F T Y T W O
```

i.d.t. – seventy

```
  T W E N T Y
  E L E V E N
  E L E V E N
  E L E V E N
  E L E V E N
    T H R E E
    T H R E E
  -----------
S E V E N T Y
```

the vanishing square

directed approach: page 97
solution: page 117

C	D	H	I	K	O	P	S	T	W
0	1	2	3	4	5	6	7	8	9

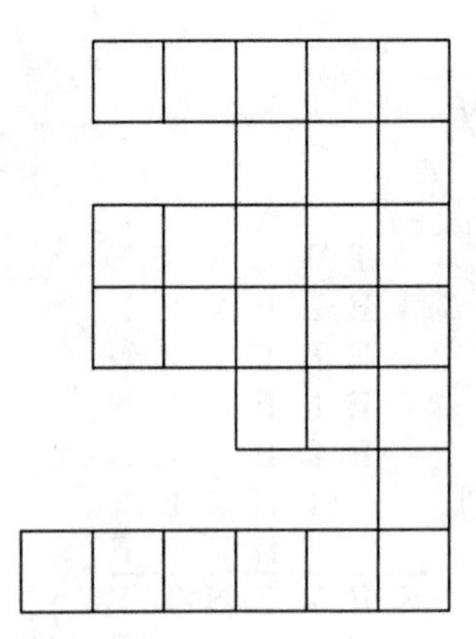

the vanishing square

I assembled sixteen identical toothpicks to form a pattern of five squares, as shown below:

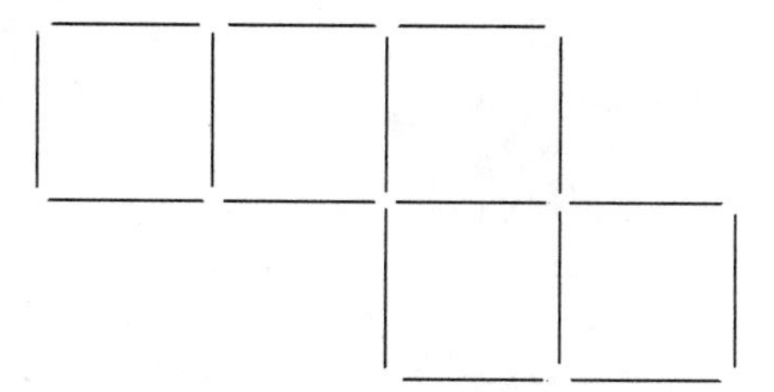

I then picked up exactly two of the toothpicks and moved them to different locations. In doing so, I formed *four* squares, each the same size as the ones in the original configuration. Moreover, every toothpick was used as the side of some square in the new arrangement.

W H I C H
T W O
T O O T H
P I C K S
D I D
I
S W I T C H ?

For maximal efficiency here, make the minimal SWITCH.

i.d.t. – ninetyone

directed approach: page 98
solution: page 114

E	F	G	H	I	N	O	R	T	Y
0	1	2	3	4	5	6	7	8	9

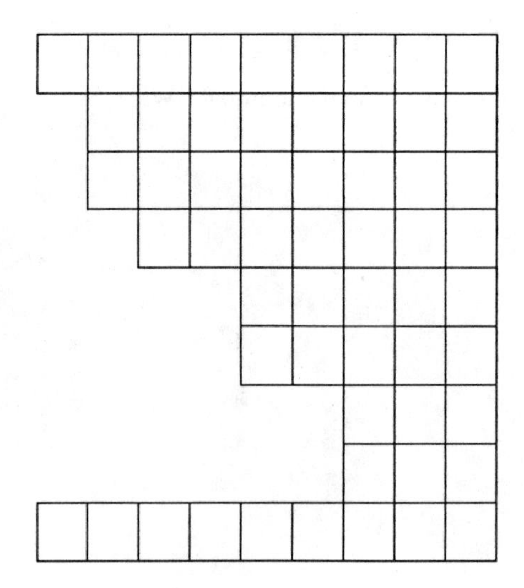

i.d.t. – ninetyone

```
T H I R T Y O N E
  N I N E T E E N
  E I G H T E E N
    F I F T E E N
        T H R E E
        T H R E E
            O N E
            O N E
-----------------
N I N E T Y O N E
```

Select any number of digits from the set {1,3,7}, where repetition is allowed. Provided that each digit is selected at least once, there is always a way to permute the digits chosen to form an integer that is divisible by 7. For example, if we select 1,1,1,1,3,3 and 7, then the integer 1,113,371 satisfies the claim. The same result holds when the original set is replaced by {3,7,9}.

high-powered equations

directed approach: page 98
solution: page 120

A	E	F	H	L	M	O	P	R	T
0	1	2	3	4	5	6	7	8	9

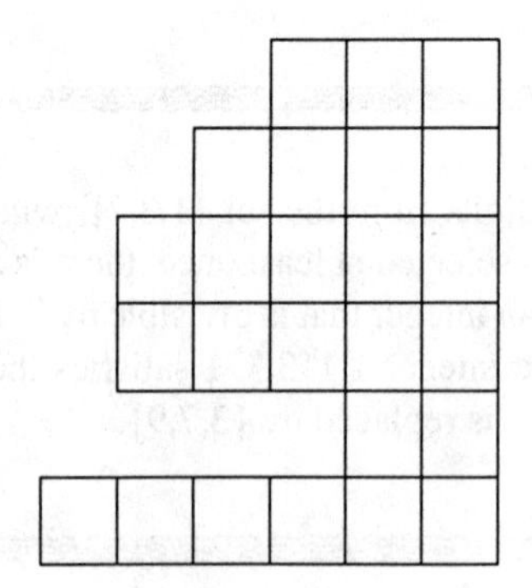

high-powered equations

Probably the most famous number problem to perplex mathematicians is a result called Fermat's Last Theorem. Pierre de Fermat (1601-1665) was undisputably the best number theorist of his day. As such, he was well aware that the Pythagorean relationship, $x^2 + y^2 = z^2$, has infinitely many solutions in which x, y, and z are positive integers. This led him to consider the existence of such solutions to the equation $x^n + y^n = z^n$ for integer exponents $n > 2$. For these n, however, his search proved fruitless. Fermat claimed to have a truly remarkable proof that integral solutions could never be found, but he excluded the details of his argument, claiming that the margin was too small to contain them.

Either Fermat had greater insight into this problem than anyone else or he was a prankster who relished a good practical joke. In any event, generations of mathematicians spent years trying to prove what Fermat indicated was eminently provable. While many "proofs" surfaced, each one was exposed as fallacious. Finally, in 1993, Andrew Wiles of Princeton University ostensibly verified the conjecture. Perhaps, at last, we now know

```
      T H E
    F O O L
  P R O O F
  P R O O F
        O F
-----------
F E R M A T.
```

i.d.t. – hundred

directed approach: page 99
solution: page 119

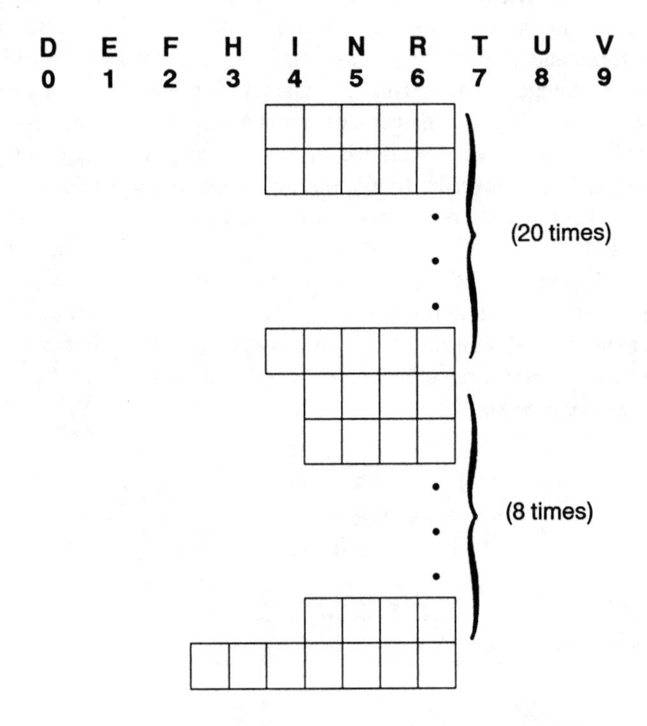

i.d.t. – eightyseven

directed approach: page 100
solution: page 117

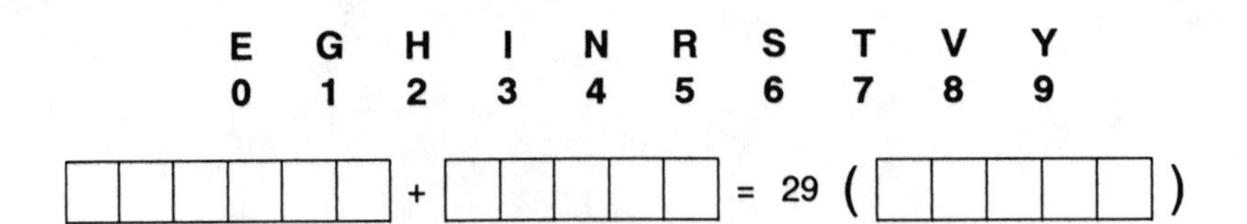

i.d.t. – hundred

```
    T H R E E \
    T H R E E |
            • |
            • } (20 times)
            • |
    T H R E E /
      F I V E \
      F I V E |
            • |
            • } (8 times)
            • |
      F I V E /
  -------------
H U N D R E D
```

i.d.t. – eightyseven

E I G H T Y + S E V E N = 29(T H R E E)
where SEVEN is divisible by 7.

sesquipedalia
directed approach: page 100
solution: page 116

D	E	H	I	N	O	R	S	T	W
0	1	2	3	4	5	6	7	8	9

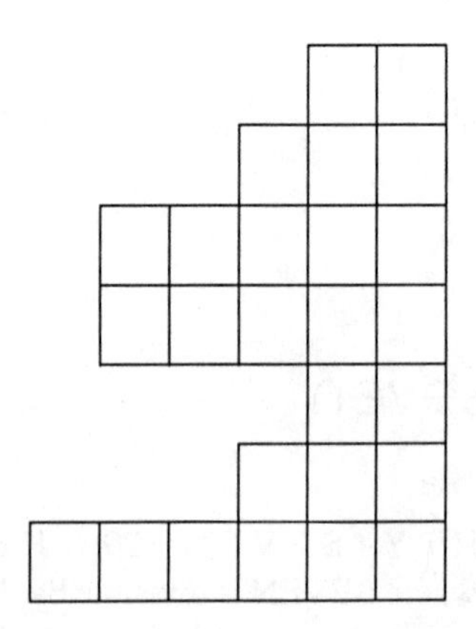

sesquipedalia

Several years ago, a colleague of mine arose at a meeting to comment upon a proposal that was under discussion. “I find this idea to be both spurious and tendentious,” he exclaimed in his usual stentorian manner. No sooner had this proclamation been made than I felt a light tap on my shoulder. Another in the audience leaned forward and asked me, in all seriousness, “Does that mean that he likes it or he doesn’t?”

```
        S O
      N O W
  T H E S E
  W O R D S
        T O
      T H E
-----------
W I S E S T
```

were inspired by and are dedicated to this pompous orator.

Presented for your consideration on page 55 are eight common proverbs, each expressed in an uncommonly complicated way. Translate them from “highfalutin’” English into everyday English so that they become recognizable. A final admonition—don’t fret if your score is not perfect. Just keep in mind that aberration is the hallmark of homo sapiens while longanimous placability and condonation are the indicia of omniscience.

To square a two-digit number that ends in a 5, add the tens digit to its square and append 25 to the sum. For example, to compute the square of 65, first find 6 + 36 = 42. Then the square of 65 is 4,225.

$$10! = (6!)\,(7!) = (3!)\,(5!)\,(7!)\,.$$

1. That prudent avis which matutinally abandons the coziness of its abode will ensnare a vermiculate creature.
2. Everything that coruscates with effulgence is not ipsofacto aurous.
3. A mass of concentrated earthly material rotating upon its axis will not accumulate an accretion of bryophytic vegetation.
4. An overabundance of mobility manufactures superfluous entities.
5. An addled dunderhead and his specie divaricate with startling prematurity.
6. One must hypesthetically exercise macrography upon that status which one will eventually tenant if one propels oneself into the troposphere.
7. Individuals who perforce are constrained to be domiciled in vitreous structures of patent frangibility should on no account employ petrous formations as projectiles.
8. A glut of talent skilled in the preparation of gastronomic concoctions will impede the quality of a certain potable solution made by immersing a gallinaceous bird in ebullient Adam's ale.

i.d.t. – sixty
directed approach: page 101
solution: page 114

E	G	H	I	N	O	S	T	X	Y
0	1	2	3	4	5	6	7	8	9

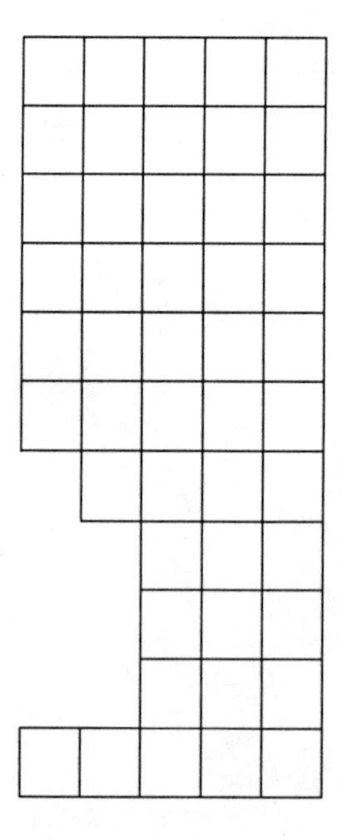

i.d.t. – sixty

```
E I G H T
E I G H T
E I G H T
E I G H T
E I G H T
E I G H T
  N I N E
    O N E
    O N E
    O N E
---------
S I X T Y
```

where EIGHT is divisible by 4.

The set {3,5,7,9,11,15,35,45,231} has the property that the sum of the reciprocals of its elements is equal to 1. There are exactly four other such sets that consist of nine positive odd integers. No set of positive odd integers that displays this property can have fewer than nine elements.

---

walking the plank

directed approach: page 102
solution: page 118

A	C	E	H	L	O	R	S	T	W
0	1	2	3	4	5	6	7	8	9

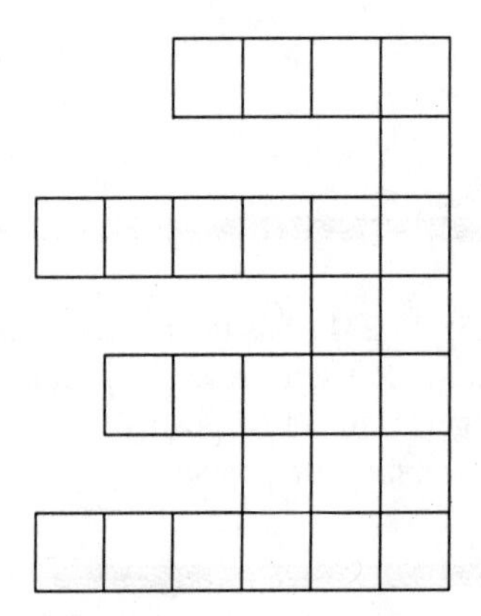

walking the plank

An ancient castle that sits upon a perfectly square island contains untold treasures worth millions and millions of dollars. To safeguard the fortune that it holds, the castle is surrounded by a 30-foot wide shark-infested moat.

Pirate Pete stands at the shoreline and longingly gazes out at the castle, dreaming of the wealth that is so near but yet so far. All that he has in his possession are two boards, each only 28.5 feet in length. Perplexed, he mutters to himself,

```
    " W H A T
            A
  H A S S L E
          T O
    R E A C H
        T H E
  -----------
  C A S T L E. "
```

But it *can* be done and should be accomplished with the least possible HASSLE. Can you use your mathematical prowess to help Pete attain his goal?

3, 4, and 5 are the only consecutive positive integers, the sum of whose cubes is itself a perfect cube.

---

it speaks for itself
directed approach: page 103
solution: page 116

A E H R S T Y

0 1 2 3 4 5 6 7 8 9

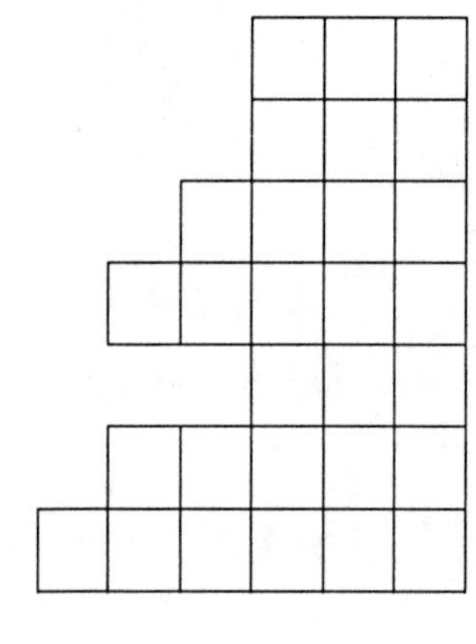

i.d.t. – forty
directed approach: page 103
solution: page 120

E F I N O R S T V Y

0 1 2 3 4 5 6 7 8 9

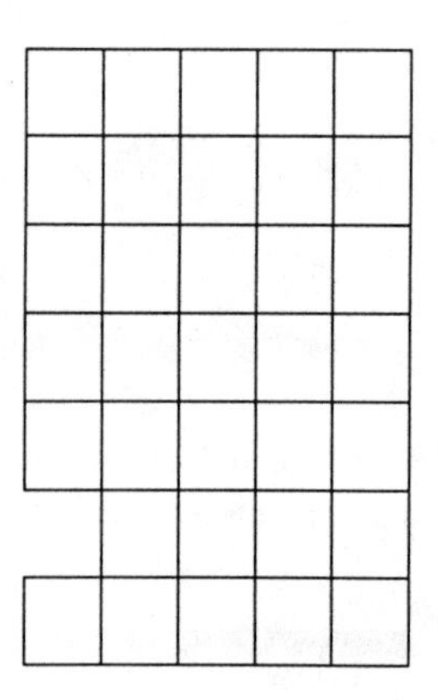

it speaks for itself

If you fancy yourself to be a good speller, then this puzzle should be right up your alley. In addition to decoding the alphametic that follows, find the spelling mistakes that it predicts and in fact contains.

```
    Y E S
    Y E S
  H E A R
T H E R E
    A R E
T H R E E
---------
E R R E R S.
```

i.d.t. – forty

```
S E V E N
S E V E N
S E V E N
S E V E N
S E V E N
  F I V E
---------
F O R T Y
```

where FIVE is divisible by 5.

cryptic clues
directed approach: page 104
solution: page 113

A	D	E	I	L	M	N	R	S	
0	1	2	3	4	5	6	7	8	9

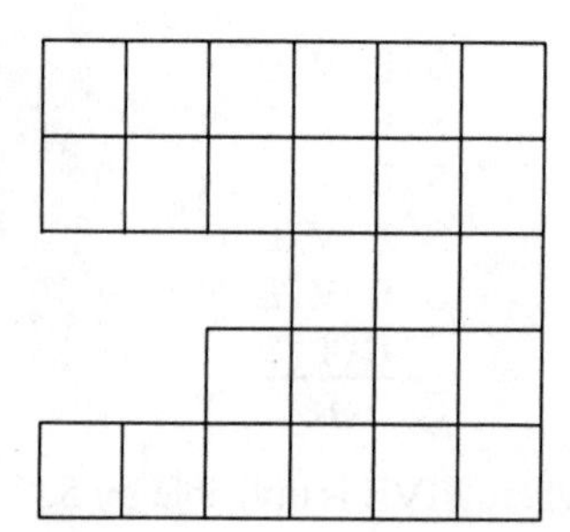

cryptic clues

Let us formally introduce

```
A M A N D A
A N D R E A
      A N D
    A N N E,
-----------
L A D I E S
```

with something in common. Each one of them was born on January 1st of a different calendar year in the twentieth century, so that each remains a given age during an entire calendar year. Armed with the following data, deduce the years that the three were born.

1. In 1990, Andrea was the age Amanda had been when Anne was twice the age Andrea was in 1989.
2. In 1991, Andrea was the age Amanda had been when Andrea was five years younger than Amanda had been when Andrea was born.
3. In 1992, Amanda was the age Anne had been when Amanda was five years younger than Anne had been when Amanda was born.

The nth repunit, denoted by R_n, is the n-digit number, each of whose digits is equal to 1. For $n \leq 1,000$, only four prime repunits are known: R_2, R_{19}, R_{23}, and R_{317}.

---

i.d.t. – fiftysix
directed approach: page 105
solution: page 114

E	F	I	N	O	S	T	V	X	Y
0	1	2	3	4	5	6	7	8	9

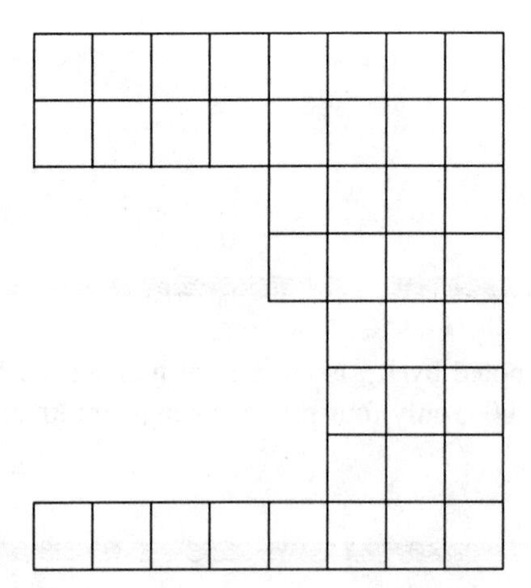

i.d.t. – fiftysix

```
N I N E T E E N
N I N E T E E N
        F I V E
        F I V E
          S I X
          O N E
          O N E
---------------
F I F T Y S I X
```

where FIFTYSIX is even.

To find the digital sum of an integer, simply add its digits and repeat the process until a one-digit sum is obtained. For example, the digital sum of 5,879 is 2, because $5 + 8 + 7 + 9 = 29$, $2 + 9 = 11$, and $1 + 1 = 2$. If p and q are twin primes (i.e., primes that differ by 2) other than the pair 3 and 5, then the digital sum of pq is always equal to 8.

---

everything's relative

directed approach: page 105
solution: page 118

A	F	I	L	M	O	R	S	T	Y
0	1	2	3	4	5	6	7	8	9

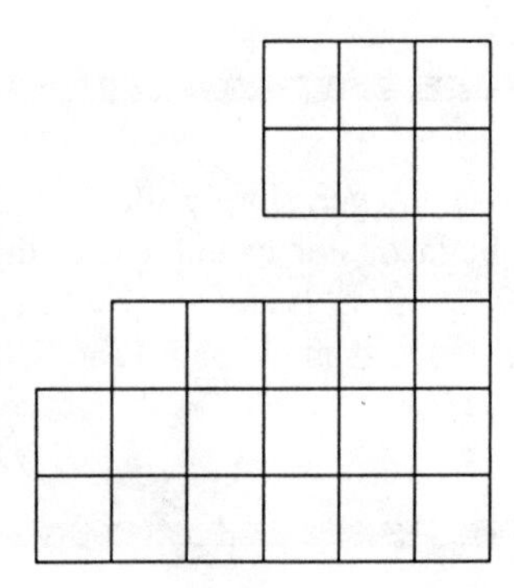

everything's relative

"We'd like a table for dinner," a restaurant patron informs the maitre d'.

```
      I T' S
      F O  R
           A
  S M A L  L
F A M I L  Y
------------
A F F A I  R.
```

"Our party consists of a mother and a father, a sister and a brother, an uncle and an aunt, a son and a daughter, a niece and a nephew, and two cousins."

"I'm terribly sorry," responds the maitre d', "but all of our larger tables are currently occupied. There will be a rather long wait before we can seat all of you."

"No, no!," exclaims the patron. "A table for four will do quite nicely, thank you."

Without putting more than one body in a chair, explain how.

3,816,547,290 is the unique pandigital integer with the property that the number formed by considering its first n digits is divisible by n for $1 \le n \le 10$.

---

i.d.t. – eighty
directed approach: page 106
solution: page 115

E	G	H	I	L	O	T	V	W	Y
0	1	2	3	4	5	6	7	8	9

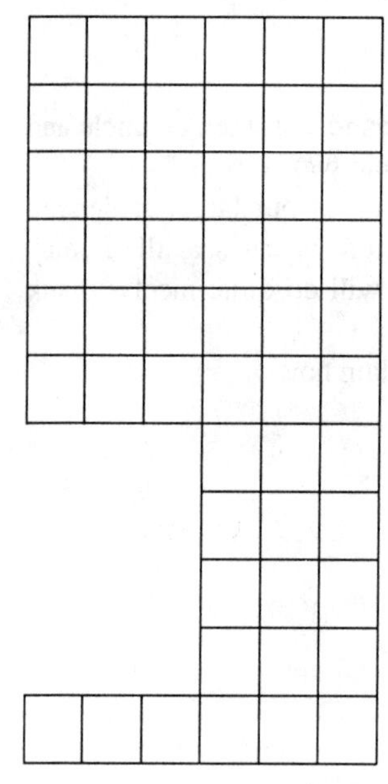

i.d.t. – thirtythree
directed approach: page 106
solution: page 119

E	H	I	L	N	O	R	T	V	Y
0	1	2	3	4	5	6	7	8	9

2 ( ) + + =

+

i.d.t. – eighty

```
T W E L V E
T W E L V E
T W E L V E
T W E L V E
T W E L V E
T W E L V E
      T W O
      T W O
      T W O
      T W O
-----------
E I G H T Y
```

i.d.t. – thirtythree

2(E L E V E N) + T E N + O N E = T H I R T Y + T H R E E

where ELEVEN is divisible by 11.

prestidigitation
directed approach: page 107
solution: page 114

D	E	H	I	N	O	R	T	U	Y
0	1	2	3	4	5	6	7	8	9

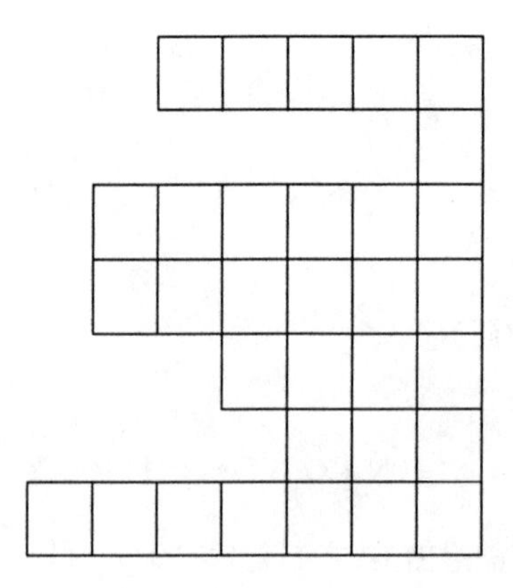

prestidigitation

"Come quickly!" my nine-year-old called to me excitedly, "I've just discovered something new in mathematics!" Preoccupied with my own unsuccessful attempts at discovery, my response was anything but swift. This lethargy prompted a second, more detailed plea:

```
      " H U R R Y !
                I
      T U R N E D
      T H I R T Y
          I N T O
            O N E
    -------------
    H U N D R E D ,"
```

he announced loudly and proudly. This proclamation definitely managed to gain my attention. As I bounded up the stairs, my first thought was to sell this secret to my stockbroker, who up to now has only been able to execute the converse transformation.

What my budding genius had done was actually some very elementary cryptography. He simply replaced each letter in the word THIRTY with its numerical positional value in the alphabet and added the resulting numbers. Thus, the "positional sum" of THIRTY became $20 + 8 + 9 + 18 + 20 + 25 = 100$, indeed backing up his claim.

The smallest positive integer solution of the equation

$$\frac{1}{x^2} + \frac{1}{y^2} = \frac{1}{z^2}$$

is $x = 20$, $y = 15$, $z = 12$.

The largest integer which can be obtained as a product of positive integers that add up to 100 is 7,412,080,755,407,364, which equals

$$(3)^{32}\,(2)^2.$$

After congratulating him on his efforts, I challenged him to extend them further by posing the following questions. I haven't seen much of my son lately. Perhaps you can offer him some assistance.

1. There is only one other integer whose positional sum is equal to 100. Which?
2. The integer TWO HUNDRED AND FIFTY NINE* has the curious property that it is equal to its positional sum. There is only one other integer with this property. Find it.
3. There is a unique integer with the property that its positional sum is equal to twice the integer. Which one?
4. Begin with the fact that FIVE + FIFTEEN = TWENTY and note that (positional sum of FIVE) + (positional sum of FIFTEEN) = 42 + 65 = 107 = (positional sum of TWENTY). Find another sum with this addition-preserving property.
5. Find an ordered set of two or more integers that displays the intriguing property of closure. That is, align the elements in the collection so that the positional sum of the first is equal to the second, the positional sum of the second is equal to the third, . . . , and the positional sum of the last is equal to the first. Such an ordered collection of integers forms a loop, where the starting point necessarily can be any integer that appears within it. Find the only such loop that exists.

* For uniformity, we will agree to use the word "AND" when appropriate if we refer to integers greater than 100.

i.d.t. – fifty
directed approach: page 107
solution: page 116

E	F	H	I	N	R	S	T	V	Y
0	1	2	3	4	5	6	7	8	9

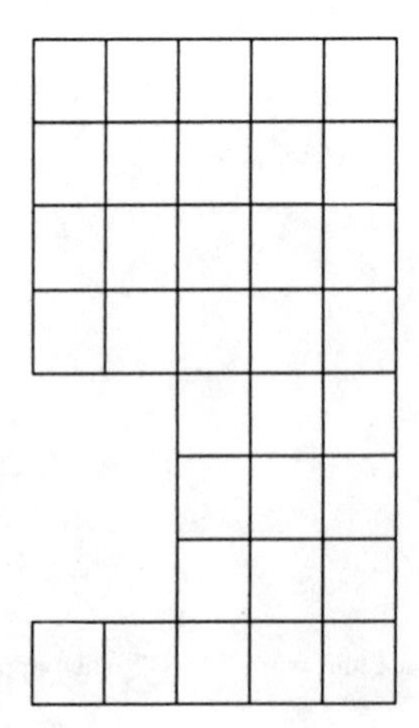

i.d.t. – fifty

```
S E V E N
S E V E N
T H R E E
T H R E E
    T E N
    T E N
    T E N
---------
F I F T Y
```

where FIFTY is divisible by 5.

By placing arithmetic symbols or no symbol at all between the digits written consecutively from 1 to 9, the resulting expression can be made to equal 100. The greatest number of symbols that can be inserted is eight:

$$1 + 2 + 3 + 4 + 5 + 6 + 7 + 8 \times 9 = 100\,.$$

The least number of inserted symbols is three:

$$123 - 45 - 67 + 89 = 100\,.$$

---

up in smoke

directed approach: page 108
solution: page 115

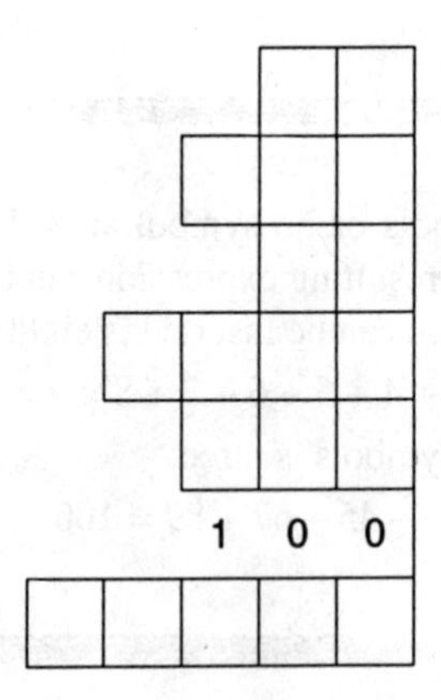

up in smoke

Harry The Hobo has scrupulously collected one hundred cigar butts. He makes no excuses for his frugality—his motto is

```
    N O
  I F S
    N O
A N D S
  B U T
  1 0 0
---------
B U T T S.
```

First, determine the unique solution to this alphametic, where the digits 0 and 1 are available for reuse. Then suppose that the tobacco from any four cigar butts can be re-rolled to form a brand-new cigar. If Harry wishes to smoke exactly three cigars per day, for how many consecutive days will his wish come true?

The smallest integral solution of

$$a^4 + b^4 + c^4 + d^4 = e^4$$

is a = 30, b = 120, c = 272, d = 315, e = 353.

---

i.d.t. – fortyone

directed approach: page 109
solution: page 118

E	F	L	N	O	R	T	U	V	Y
0	1	2	3	4	5	6	7	8	9

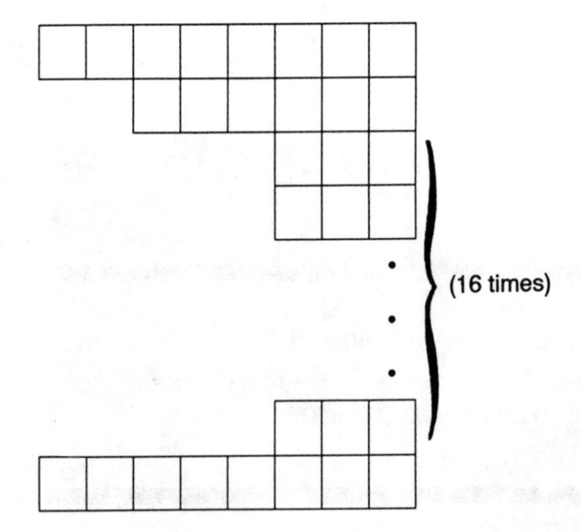

i.d.t. – fortyone

```
F O U R T E E N
      E L E V E N
            O N E \
            O N E |
                • |
                • }  (16 times)
                • |
            O N E /
―――――――――――――――
F O R T Y O N E
```

No positive integer starting with 9 and having all nonzero digits arranged in consecutive descending order (e.g., 987,654,321,987) can be a prime. There are primes that start with 1 and have all nonzero digits arranged in consecutive ascending order, the smallest being 1,234,567,891.

an inviting problem
directed approach: page 109
solution: page 113

A	C	E	M	O	P	R	T	U	Y
0	1	2	3	4	5	6	7	8	9

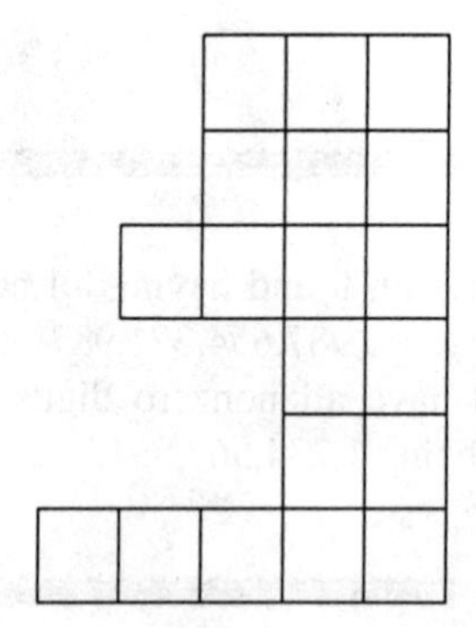

an inviting problem

I'm planning an elaborate dinner party, but prospective guests are not expected to arrive empty-handed. In order to gain entrance, each must bring along an appropriately selected item. A dozen roses would not be acceptable, but a dozen daffodils would be fine. Some cookies would be appreciated, but a fruitcake would not be. You'll surely get in with a bottle of chardonnay, but not with a bottle of chablis. Ice cream is fine as long as it's vanilla; chocolate simply will not do.

If this type of gala seems intriguing, perhaps you'd like to join me? All that you have to do is figure out how to "bring the right thing." If you can, then

```
  Y O U
  M A Y
C O M E
    T O
    M Y
-------
P A R T Y,
```

where an odd PARTY it certainly is!

? ! ? ! ? ! ? ! ? ! ? ! ? ! ? ! ? ! ? ! ?

SECTION 2

DIRECTED APPROACHES

? ! ? ! ? ! ? ! ? ! ? ! ? ! ? ! ? ! ? ! ?

cover

```
        A T
      L A S T
  E N C O D E D
    T O T A L S
    S E C O N D
---------------
A D D I T I O N
```

The value for A is immediately determined from the leftmost column. Once this is done, the value for D is available from the millions column, which also offers only two possibilities for E. One of these is eliminated by considering the value of the sum N + T + S from the hundred thousands column in conjunction with the equation involving these same letters that comes from the units column. The value for E is thus discovered. Now reconsider the previous two sources (there are two possible values for N + T + S) and specifically assign values for N, T, and S wherever appropriate digits are present. Pursuing each existing avenue, examine the tens column to find values for L and O. In only one instance will the hundred column yield an acceptable value for I. Continuing leftward, the thousands column produces the remaining digit as the value for C, and the ten thousands column behaves properly to confirm the solution.

dedication

```
          Y E S
    T H E Y'R E
T E R R I F I C
---------------
T E A C H E R S
```

The units column establishes the relationship between E and C, which in turn leads to the relationship between E and I from the tens column. The possible carryover into the thousands column gives potential assignments for H, one of which can be immediately eliminated because it implies that R = I. For each of the others, account for carryovers and determine specific equations from the hundreds, ten thousands, and hundred thousands columns. Consider these equations together with those found from the units and tens columns and list available values for E. Then, as far as possible in each instance, obtain the exact values for C (units column), I (tens column), R (ten thousands column), Y and F (hundreds column), and T and A (hundred thousands column). In the single case where this pursuit is successful, assign the remaining digit to S.

even odds

```
      A
  FIFTY
  FIFTY
 ------
 CHANCE
```

The value for C is immediate from the leftmost column. Since the digit zero is not available, the tens column offers only one possibility for T. From the ten thousands column, there are limited options for F. In each instance, use the hundreds column to assign N where feasible, and then use the ten thousands column to establish H. Next, obtain the equation involving I and A from the thousands column, and satisfy it with the digits that remain. Once A is known, shift to the units column to relate Y and E, and conclude by determining their specific values.

go for the gold

```
     IN
  WHICH
     IS
    THE
  LIGHT
 ------
 WEIGHT
```

From the hundred thousands, ten thousands, and thousands columns come definite values for W, L, E, and H. The units column then gives the sum N + S, with only two digit pairs available to achieve it. Keeping these options in mind, an explicit equation follows from the hundreds column which yields the values for C and I. The assignment for I produces T from the hundreds column, while the choice for C establishes a unique interchangeable pair of digits for N and S. The only digit that remains becomes the value for G.

i.d.t. – ninety

```
   S I X T Y
   E I G H T
   T H R E E
     N I N E
       T E N
 -----------
 N I N E T Y
```

The leftmost column permits only two possible values for N. For each, shift to the units column to determine the parity of T. List the options for T and then use the relationship from the units column to find E. Next, obtain the value for H from the tens column and use it to establish the value for I from the thousands column. Moving leftward, the ten thousands column gives the value for S. In any remaining case, the hundreds column yields an explicit value for the sum X + G + R. Satisfy it with a trio of interchangeable digits wherever availability allows. Finally, the only unassigned digit is associated with Y.

i.d.t. – twenty

```
 E L E V E N
   T H R E E
   T H R E E
       O N E
       O N E
       O N E
 -----------
 T W E N T Y
```

Establish the parity of E from the units column and noting its presence in the leftmost column, list all possibilities. Examine the carryover into that column to obtain values for T. In each instance, find the value for N from the tens column, making certain that that value yields the appropriate carryover into the tens column. Whenever this is so, fix the value for Y from the units column. Next, the thousands column offers limited options for H. Each of these in turn leads to a specific equation linking L and W when the ten thousands column is considered. Satisfy this equation with available digits. Finally, the hundreds column produces a relationship that involves V, R, and O which can be made true in just one case.

wholes with holes

```
  F R O M
  Y O U R
  F O U R
  F O R M
    O U R
  F O U R
---------
F O R M S
```

View the thousands column and the ten thousands column simultaneously to find the value for F. Considering the thousands column again and tracking the carryover from the hundreds column yields possible assignments for the pair Y and O. The carryover into the hundreds column restricts U to just two possible values, eliminating two cases. For the ones that remain, set up equations from the units and tens columns. In only one instance will the remaining digits be available for R, M, and S.

i.d.t. – sixtyone

```
N I N E T E E N
T H I R T E E N
    S I X T E E N
        T H R E E
        T H R E E
            O N E
            O N E
            O N E
            O N E
            O N E
            O N E
            O N E
-----------------
S I X T Y O N E
```

The units and tens columns jointly produce choices for N and E, with the presence of N in the leftmost column somewhat limiting the options. This same column also offers assignments for the pair T,S in each instance, where the millions column subsequently yields possibilities for the value for H. From the hundreds column, establish a relationship between R and O, and satisfy it with available digits when this can be done. Move next to the thousands column to find the value for Y and continue leftward to obtain the value for X from the ten thousands column. In only one case will the remaining digit be compatible with the equation involving I that results from the hundred thousands column.

division decisions

```
     W E' V E
D E C I D E D
  B E F O R E
          W E
      E V E R
-------------
D I V I D E D
```

Start by listing the digit pairs that are appropriate for the relationship between E and R obtained from the units column. Then proceed leftward, finding the values for the sums V + W, O + V, and F + W from the tens, hundreds, and thousands columns, respectively. Assign available digits to V, W, O, and F wherever possible. The ten thousands column next yields the value for C, followed by an equation involving B and I from the hundreds thousands column. Whenever this equation can be satisfied, the remaining digit becomes the value for D. Lastly, apply the divisibility test for 7 to choose the designated solution.

i.d.t. – fiftyone

```
E I G H T E E N
E I G H T E E N
      T H R E E
      T H R E E
        N I N E
---------------
F I F T Y O N E
```

The leftmost column imposes a ceiling upon the value for E. Keeping this in mind, note that the tens column makes the selection for E explicit, with the exact value for N then following from the units column. The millions column offers two possible assignments for I, each of which leads to a unique value for F from the leftmost column. Use the hundred thousands column next to establish the appropriate choice for G. From the initial condition and the previously-determined value for N, there is only one feasible carryover into the thousands column. When the hundreds column is examined, this observation yields just one set of values for the pair R,O in each existing case. The thousands column provides a relationship involving T, H, and Y which serves to eliminate one avenue completely. The remaining option permits Y to be specified. Moreover, the known value of T + H gives T from the ten thousands column and thereby gives H in retrospect. To complete the process, verify that the carryover into the hundred thousands column is the correct one.

optical allusion

```
  T W E N T Y
  T W E N T Y
  -----------
  V I S I O N
```

The leftmost column limits the size of T, enabling the carryover into the hundreds column to be found. This gives the parity of I, while the units column establishes the parity of N. Using these observations, generate possible combinations for I and N. Refer to the ten thousands column to find the value for W. Then reexamine the leftmost column and make assignments for T and V where feasible. With V known, the tens column produces the value for O. A second visit to the units column yields the value for Y. Finally, the thousands column relates E and S, with appropriate digits available in only one case.

open and shut case

```
      O P E N
    D O O R S
  C L O S E D
  C L O S E D
    D O O R S
  -----------
  O P E N E D
```

The given condition suggests an appropriate first choice for O. This quickly leads to an explicit carryover into the ten thousands column, a determination of the parity of P, and two possibilities for C. For each of these, examine the equation that results from the ten thousands column and list the various available assignments for the trio P, D, and L. Note the parity of the carryover into the tens column next, and find its exact value by looking at the units column. This yields an equation involving D, N, and S. For each previously-listed value for D, assign unused digits for N and S wherever possible. If the hundreds column is a feasible result, the carryover into the thousands column produces the value for E. Lastly, appeal to the tens column to establish the value for R.

i.d.t. – thousand

$$5(HUNDRED) + 10(TEN) + 400(ONE) = THOUSAND$$

Begin by making one ten-shift and four hundred-shifts, thus converting the problem into

```
    H U N D R E D
    H U N D R E D
    H U N D R E D
    H U N D R E D
    H U N D R E D
            T E N
          O N E
          O N E
          O N E
          O N E
  ---------------
  T H O U S A N D
```

Considering the carryover into the tens column, the value for D is immediately apparent from the units column. Use the two leftmost columns to establish possible assignments for H and T, with the hundred thousands column next yielding choices for U and O. Then move to the right to obtain the value for N from the ten thousands column. List appropriate digits for E from those that are still available, noting that the tens column fixes the parity of E. For each such choice, the hundreds column offers values for A, which in turn leads to values for R. In only one instance will the thousands column produce the single remaining digit for S, where the resulting carryover into the ten thousands column is the appropriate one.

urban affairs

```
   L E T' S
   L I S T
       A T
 L E A S T
-----------
T H I R T Y
```

The values for T and H are apparent from the two leftmost columns. Examining the carryover into the ten thousands column eliminates one of the two possibilities for L, yielding its value as well. The carryover into the tens column can next be uniquely determined. Then consider the equations that are generated by the units and tens columns. These give possible assignments for the letters S, Y, and A. In each case, find the relationship among E, I, and R from the hundreds column. From the subsequent carryover into the thousands column, find possible values for E and I. Then backtrack to obtain R, which can be accomplished in precisely one of the cases.

i.d.t. – thirty

```
  S E V E N
  T H R E E
      F I V E
      F I V E
      F I V E
      F I V E
-----------
T H I R T Y
```

The leftmost column immediately provides the value for T, while the units column establishes the parity of E. List the possibilities for E and use the known parity of the carryover into the tens column to obtain V from that column. The assignment for I follows next from the hundreds column. Observing from the ten thousands column that the value for T places a constraint upon H, inspect the thousands column to find the relationship between H and F. Satisfy it wherever available digits allow, and subsequently get S from the ten thousands column. Revisit the units column to catalogue acceptable choices for N and Y, keeping the required carryover into the tens column in mind. This is successful in only one case. The remaining digit is finally associated with R.

splitting hares

```
           S O
             I
     T O A S T
           T O
       F A S T
     B I R T H
   H A B I T S
           O F
     P A I R S
           O F
 -------------
 R A B B I T S
```

The three leftmost columns produce the values for R, A, and H at the start. Next, find the exact carryover into the thousands column. This same column then establishes two possible values of the sum O + F + I, with the ten thousands column yielding a unique value of the sum T + P in each instance. One of the options for the sum O + F + I offers only one assignment for the trio O, F, I, but this makes T + P unattainable. By the process of elimination, specific values for both O + F + I and T + P are discovered. Looking at the digits still available gives the minimum possible carryover into the hundreds column. This column suggests a maximum value for I, while the previously-found value of O + F + I gives a minimum value for I. Together, these facts limit the choices for I. For each, the pair O,F can be uniquely determined; in all but one case, the remaining digits are not adequate to form the sum T + P. Thus, the value for I is found, and with it, values for the pair O,F. Using this data, the units column relates T and S, as well as giving the parity of S. Reconsidering the tens column, the parity of S produces the parity of the carryover into the tens column, which in turn establishes the exact value of this carryover. Since the carryover into the hundreds column is known, specific equations are at hand from the two rightmost columns. Only one value for S leads to acceptable values for T and O. Referring to earlier-obtained equations, these generate values for P and F, respectively. The sole remaining digit is finally assigned as the value for B.

i.d.t. – fortytwo

FOURTEEN + 2(THREE) + 11(TWO) = FORTYTWO

Use a ten-shift to convert the given problem into

```
FOURTEEN
    THREE
    THREE
      TWO
       TWO
FORTYTWO
```

Then the units column yields a relationship between E and N. After listing possible assignments for this pair, examine the tens column in each case to determine the value for O, remembering to heed the initial condition. Proceed leftward to obtain appropriate choices for R and W from the hundreds column. Once R is fixed, go to the hundred thousands and ten thousands columns to discover the values for U and T, respectively. Wherever remaining digits permit, the thousands column next provides values for H and Y. The only unused digit becomes the value associated with F.

for the cruciverbalists

```
    FOR
    ONE
    ODD
    ODD
  CROSS
   WORD
 PERSON
```

The assignments for P, C, and E are immediate from the two leftmost columns. Shifting attention to the two rightmost columns, establish equations from them. Consider these simultaneously to obtain a relationship involving N, D, and the carryover into the tens column. For each possible value of this carryover, that relationship yields the digits for N and D. Then the tens column produces possibilities for R and S, each of which generates values for O and F when the hundreds column is examined. The remaining digit is appropriate for W in exactly one instance.

i.d.t. – sixtysix

```
T H I R T E E N
T H I R T E E N
T H I R T E E N
T H I R T E E N
        N I N E
          O N E
          O N E
          O N E
          O N E
          O N E
---------------
S I X T Y S I X
```

The leftmost column permits only two possible selections for T. For each one, examine the ten thousands column to find acceptable choices for R. The hundred thousands column next provides assignments for the pair I,X, where the parity of the carryover into the millions column must match the parity of I. This being the case, use the millions column to obtain the value for H. Follow this by determining the value for S from the leftmost column. Then shift to the units column to list available digits for the pair N,E, keeping in mind that the parity of the carryover into the tens column and the parity of I must agree. Whenever the equation from the tens column is satisfied, go to the hundreds column to produce the value for O. Lastly, check the thousands column to find the one existing case in which Y is equal to the only remaining digit, and confirm that the carryover into the ten thousands column is the appropriate one.

count on it!

```
    T R U E
      O U R
    L I S T
        I S
  Q U I T E
-----------
U N I Q U E
```

Begin by noting the values for U, N, and Q from the leftmost pair of columns. Shift to the right in order to establish equations from the units, tens, and hundreds columns. Employing the remaining digits, find limitations on the size of the sum S + T. In each case, next determine the value of R + E from the units column and the value for I from the tens column. Cataloguing the possibilities for R and E, use the hundreds column to assign the value for O. Then do the same for the available choices for S and T, and use the thousands column to get L.

i.d.t. – fiftytwo

```
    S I X T E E N
    S I X T E E N
          N I N E
          N I N E
            O N E
            O N E
  ---------------
  F I F T Y T W O
```

Use the three leftmost columns to find conceivable values for the trio F, I, and S. Examine the ten thousands column next and list the digits that can be assigned to X. Keeping in mind the carryover into the ten thousands column, obtain values for T from that column. Then follow this by determining the possibilities for N and Y from the thousands column. With N fixed, the units column relates E and O. Wherever this relationship can be satisfied, move to the tens column to see if the remaining digit is the appropriate one for W. Be sure to check the carryover into the hundreds column for correctness.

i.d.t. – seventy

```
  T W E N T Y
  E L E V E N
  E L E V E N
  E L E V E N
  E L E V E N
    T H R E E
    T H R E E
-------------
S E V E N T Y
```

Noting that the parity of the carryovers into the hundreds and ten columns are available, find the only possible values for E and N from the tens column and units column, respectively. Then determine the parity of the carryover into the thousands column and establish the explicit choices for this carryover. Each of them leads to a unique value for H, with the value for S following next from the carryover into the leftmost column. In each case, return to the hundreds column to write an equation involving V and R, and list pairs of digits that satisfy it. Then move to the ten thousands column to obtain a relationship involving W, L, and T, observing that the hundred thousands column limits the options for T. Only once are appropriate digits still present for these assignments, leaving the remaining digit as the correspondent for Y.

the vanishing square

```
  W H I C H
      T W O
  T O O T H
  P I C K S
      D I D
          I
-----------
S W I T C H
```

Determine the value for S from the leftmost column. Considering the carryover into the ten thousands column, establish the value of the sum T + P. Then reexamine the ten thousands column to find possible values of the sum H + O. For each of these, go to the units column to find the value of the sum D + I. Next, obtain potential values of W + T + K + I from the tens column, and follow this by obtaining the value of O + C from the hundreds column. For each available O, compute the values for both H and C. Then list digit pairs for T and P, as well as for D and I. At this juncture, only two digits remain, so their sum must be W + K. Take this information back to the tens column to yield the sum T + I and satisfy this relationship wherever feasible. Specific assignments for T, I, P, and D can now be made, leaving the values for W and K interchangeable. Conclude by selecting from all solutions the one that displays the minimal value for W.

i.d.t. – ninetyone

```
 T H I R T Y O N E
   N I N E T E E N
   E I G H T E E N
     F I F T E E N
         T H R E E
         T H R E E
             O N E
             O N E
 -----------------
 N I N E T Y O N E
```

View the units and tens columns together to find a limited number of possibilities for the values of N and E. For each N, examine the leftmost column to determine T. Noting the required parity of the carryover into the thousands column, use the hundreds column to obtain values for the sum R + O and list viable options for R and O. The thousands column next yields the value for H, followed by the assignment for F from the ten thousands column. For each R, move leftward to establish the presence of any available digits for the letters G and I. Once I has been selected, check the truth of the equations that result from the millions and ten millions columns. This will prove successful in only one case. Conclude by assigning the single remaining digit to be the value for Y.

high-powered equations

```
      T H E
    F O O L
  P R O O F
  P R O O F
        O F
 -----------
F E R M A T
```

Begin by finding the value for F from the leftmost column. Observe next that the carryover into the units column can be explicitly determined, leading to possible values for the pair P and E. In each of these cases, establish an equation from the units column involving L and T, and list available digits that satisfy it wherever possible. The thousands column offers limited options for R. For each remaining avenue, consider appropriate values for O that would generate the correct carry-over into the thousands column. Finally, inspect the tens column to obtain values for H and A, with the carryover into the hundreds column then yielding the value for M.

i.d.t. – hundred

$$20(THREE) + 8(FIVE) = HUNDRED$$

Perform two ten-shifts to transform the problem into

$$\begin{array}{ccccccc}
 & T & H & R & E & E & \\
 & T & H & R & E & E & \\
 & & & F & I & V & E \\
 & & & F & I & V & E \\
 & & & F & I & V & E \\
 & & & F & I & V & E \\
 & & & F & I & V & E \\
 & & & F & I & V & E \\
 & & & F & I & V & E \\
 & & & F & I & V & E \\
\hline
H & U & N & D & R & E & D
\end{array}$$

Begin by observing the value for H from the leftmost column. Go next to the units column and list selections for E and D. Eliminate half of these options by noting that the carryover into the tens column must match the parity of E. For those that remain, obtain V from the tens column and use the resulting carryover to establish a relationship between I and R from the hundreds column. Satisfy it wherever possible, keeping in mind that the value for D fixes the parity of the carryover into the thousands column. This column yields the value for F, with the subsequent carryover into the ten thousands column then giving N. The equation involving T and U found from the hundred thousands column then admits the two remaining digits as a solution in exactly one case.

i.d.t. – eightyseven

$$\text{EIGHTY} + \text{SEVEN} = 29(\text{THREE})$$

Add THREE to both sides of the given equation, use three ten-shifts, and consider the equivalent problem presented below:

```
E I G H T Y        T H R E E –
  S E V E N        T H R E E –
  T H R E E    =   T H R E E –
___________        ___________
                             0
```

From the tens columns, establish the possible relationships between T and E. Then examine the leftmost columns to list specific assignments for T and E. In each instance, the units columns lead to values for the sum Y + N. Associate interchangeable digits with this pair of letters as available. Pass next to the hundreds columns and determine the exact value of the sum H + V + R, using the ten thousands column of the second sum to limit the choices for H. For each such choice, find feasible values for the pair V,R. Moving to the left, check the thousands columns to see if any selection for R generates an unassigned value for G. When this is so, the two remaining digits must be interchangeable values for the pair S,I. Verify the accuracy of the results from the ten thousands columns. Then select the single solution that satisfies the initial condition.

sesquipedalia

```
        S O
      N O W
  T H E S E
  W O R D S
        T O
      T H E
___________
W I S E S T
```

The value for W is evident from the leftmost column, with the value for I following immediately after from the ten thousands column. This same column also offers two possibilities for T, both of which yield the same carryover into the thousands column. For each choice for T, examination of the units column determines the parity of S. Each acceptable value for S leads to the value of the sum H + O from the thousands column. Assign digits to H and O as available. Then reconsider the units column to find the value for E, and move leftward to obtain the value for D from the tens column. Finally, the hundreds column gives an interchangeable pair of digits for N and R in exactly one instance.

i.d.t. – sixty

```
E I G H T
E I G H T
E I G H T
E I G H T
E I G H T
E I G H T
  N I N E
    O N E
    O N E
    O N E
---------
S I X T Y
```

The leftmost column gives the value for E at the outset. Note that the initial condition yields the parity of T, which must match the parity of the carryover into the tens column. Then use the units column to list possible assignments for T and Y. Once again employ the initial condition, this time to establish the parity of H. Select H accordingly and for each choice, generate N from the tens column. Then pass to the thousands column to obtain alternatives for I, with the resulting carryover into the ten thousands column producing the value for S. Finally, examine the hundreds column to get an explicit equation involving G, O, and X, and satisfy it with the digits that remain.

walking the plank

```
     W H A T
           A
 H A S S L E
         T O
   R E A C H
       T H E
 -----------
 C A S T L E
```

The ten thousands column gives the value for R at the outset, with the relationship between H and C then following from the leftmost column. Use the units column to next establish three possible values for the sum T + A + E + O + H. The smallest of these offers only one set of digits, with the value for H specified. This in turn provides the assignment for C, but the resulting equation from the tens column becomes impossible to satisfy. The largest of these possibilities again offers a unique set of digits. No digit in this set is an appropriate choice for H, however, in light of the equation involving H and C. By elimination, the exact value for T + A + E + O + H is secured. Then from the tens column, obtain a relationship involving A, T, C, and H. Recall that the problem suggests the smallest value for H and choose accordingly. From the selected value for H, determine the corresponding value for C. Then backtrack to fix the values for A + T and E + O from the tens and units columns, respectively. Moving leftward, find values for A + S from the hundreds column and for W + E from the thousands column. Assign compatible digits to these letters as available, noting again the constraint when selecting L.

it speaks for itself

```
    Y E S
    Y E S
  H E A R
T H E R E
    A R E
T H R E E
---------
E R R E R S
```

When the leftmost column is inspected, only two possible assignments are available for E. In each case, consider appropriate values for R from the ten thousands column. Go next to the units column to get the value for S, and subsequently pass to the tens column to obtain the value for A. Continuing leftward, the hundreds column yields the value for Y. Since R is at this point already known, discover values for H and T from the thousands and ten thousands columns, respectively.

i.d.t. – forty

```
S E V E N
S E V E N
S E V E N
S E V E N
S E V E N
  F I V E
---------
F O R T Y
```

The leftmost column provides just one choice for S. Further, the initial condition admits only two feasible values for E. Each of these values serves to establish the parity of N when the units column is examined, and this observation leads to the assignment for Y. Considering the carryover into the leftmost column, find the value for F, thereby eliminating one of the two existing cases. In the one that remains, list the possibilities for N and for each, use the resulting carryover into the tens column to relate V and T. Wherever this relationship can be satisfied, move to the hundreds column to get a condition involving I and R. Only once are digits still available that both meet this criterion and make the ensuing carryover into the thousands column appropriate for the selection of the value for O.

cryptic clues

```
  A M A N D A
  A N D R E A
        A N D
      A N N E
  -----------
  L A D I E S
```

Consider the thousands column in conjunction with the hundred thousands column to determine the value for A. Find two options for the sum M + N from the ten thousands column, each of which leads to a unique value for L. One of these possibilities is quickly eliminated because it produces an inadequate carryover into the thousands column. Thus obtain explicit values for L as well as for the sum M + N. The carryover into the hundreds column is the same for each pair of digits available for M and N. This yields an equation involving N, R, and I, and said equation disposes of one of the existing cases. Two others are also seen to be impossible because the tens column would require a previously-used value for D in each instance. Hence, the values for M and N become known; the latter implies the value for D, since the carryover into the tens column can be specified. The units column next establishes the relationship between E and S, and the hundreds column does the same for R and I. The pool of remaining digits leads to a unique assignment for these four letters.

i.d.t. – fiftysix

```
N I N E T E E N
N I N E T E E N
        F I V E
        F I V E
          S I X
          O N E
          O N E
---------------
F I F T Y S I X
```

The three leftmost columns give the value for I and offer limited options for the values assigned to N and F. Furthermore, consideration of the carryover into the hundred thousands column places an upper bound on the value for E. Using these observations, examine the units column next, noting that the parity of the carry-over into the tens column is predictable. This eliminates all but one existing avenue and secures the exact value for E. Then look at the ten thousands column to determine the value for T. The relationship obtained from the tens column, in conjunction with the required parity of the carryover into the hundreds column, fixes the value for V. The hundreds column then yields two possibilities for O, only one of which produces an available digit for Y when the thousands column is checked. To conclude, apply the initial condition to select the value for X, leaving the remaining digit to be the correspondent of S.

everything's relative

```
      I T' S
      F O  R
           A
  S M A L  L
F A M I L  Y
------------
A F F A I  R
```

The leftmost column establishes the relationship between F and A. The ten thousands column then offers acceptable values for S. In each instance, list possibilities for M. Noting that there must be a carryover into the thousands column, find values for F and A corresponding to each choice for M. Examine the hundreds column next to find options for I. Then shift to the units column to discover the value of the sum L + Y, and assign values to these letters as available. The tens column yields values for T and O, and the remaining unused digit is necessarily matched with R.

i.d.t. – eighty

```
T W E L V E
T W E L V E
T W E L V E
T W E L V E
T W E L V E
T W E L V E
      T W O
      T W O
      T W O
      T W O
-----------
E I G H T Y
```

The leftmost column offers an immediate assignment for T, which in turn yields the parity of the carryover into the tens column. Returning to the leftmost column, list options for E and then inspect the ten thousands column to obtain possible values for W. Look next at the feasible carryovers into the ten thousands column to find choices for I. Shift to the units column to relate O and Y. Recalling the known parity of the carryover into the tens column, satisfy that relationship wherever available digits allow. Continuing leftward, the tens column gives the value for V and establishes the carryover into the hundreds column as well. This column provides selections for L and H. In only one case does the resulting carryover into the thousands column turn out to be the appropriate one in order to match the remaining digit with G.

i.d.t. – thirtythree

$$2(ELEVEN)+TEN+ONE=THIRTY+THREE$$

The leftmost columns offer possible assignments for E and T. In each instance, consider the tens columns to obtain the value for N, followed by the determination of the value for Y from the units columns. Shift to the ten thousands columns to list the options for L and H. Then the thousands columns yield available digits for I. This being successfully accomplished, compare the hundreds columns and satisfy the resulting equation wherever possible with values for V, R, and O, remembering to account for appropriate carryovers into the thousands columns. From the set of all solutions, select the unique one that validates the initial condition.

prestidigitation

```
    H U R R Y
            I
  T U R N E D
  T H I R T Y
      I N T O
        O N E
-------------
H U N D R E D
```

The leftmost column establishes the value for H at the outset. Keeping in mind the carryover into the hundred thousands column, use that column to catalogue the options for T and U. For each, examine the ten thousands column and generate possible values for N. Next, determine the feasible carryovers into the hundreds column and use that information to find the size of the sum R + O. Assign specific values to these letters, with the choice for R made in conjunction with its appearance in the tens column. In the surviving cases, pass to the units column and obtain an equation involving Y, I, and E, satisfying it whenever available digits exist. Finally, check the thousands column to obtain the value for D.

i.d.t. – fifty

```
S E V E N
S E V E N
T H R E E
T H R E E
    T E N
    T E N
    T E N
---------
F I F T Y
```

The initial column allows only two possible assignments for Y. From the units column, each yields the value for E, as well as the parity of N. List the options for N and use the carryover into the tens column to find T. The presence of T in the leftmost column limits its size and thereby eliminates one of the original avenues from further consideration. In each of the cases that remain, use the leftmost column to establish the value for S. Learn the parity of F from the hundreds column, and reexamine the leftmost column to find F explicitly. Return to the hundreds column to obtain interchangeable digit pairs for V and R where available. Finally, satisfy the equation for H and I that results from the thousands column.

up in smoke

```
   N O
 I F S
   N O
A N D S
  B U T
  1 0 0
---------
B U T T S
```

Explicit values for B, U, and A can be found quickly from the thousands and ten thousands columns. Move next to the units column to establish the exact relationship involving O, S, and T, and list all available trios of digits that satisfy it. At this point in the process, four digits remain unassigned. Selecting each of these for I leads to the determination of the set of values for N, F, and D, hence for the sum N + F + D. Use this information together with the known carryover into the tens column to relate N and T. Whenever there is an appropriate value for N in the above-mentioned set, the remaining two set elements form an interchangeable pair for the letters F and D. Finally, check the accuracy of the hundreds column, remembering to account for the carryover into that column. In only one instance will a correct statement emerge.

i.d.t. – fortyone

FOURTEEN + ELEVEN + 16(ONE) = FORTYONE

Make a ten-shift to convert the problem into

```
FOURTEEN
   ELEVEN
     ONE
      ONE
      ONE
      ONE
      ONE
      ONE
      ONE
--------
FORTYONE
```

Then the units and tens columns together lead to specific assignments for E and N. The hundreds column next produces possible values for V and O. For each such pair, examine the thousands column to discover the relationship between T and Y, and list available digits that satisfy it. Moving to the ten thousands column, establish the value for the sum R + L and search the remaining digits to see if it can be attained. Whenever this is the case, continue leftward to get U from the hundred thousands column, thus pinpointing R and, therefore, L. Finally, assign the only unused digit to F.

an inviting problem

```
  YOU
  MAY
 COME
   TO
   MY
-----
PARTY
```

First find P from the leftmost column and A from the thousands column, noting that there are but two available values for C. Using this last observation, the units column establishes the exact value of the sum U + E + O + Y, hence the carryover into the tens column. This column next gives a relationship between O and M. For each digit pair that satisfies it, an examination of the hundreds column gives C explicitly. The same source yields possible assignments for Y and R, where Y is restricted to be odd from the initial condition. Return to the units column once again to find interchangeable values for U and E, and match the one remaining digit with T.

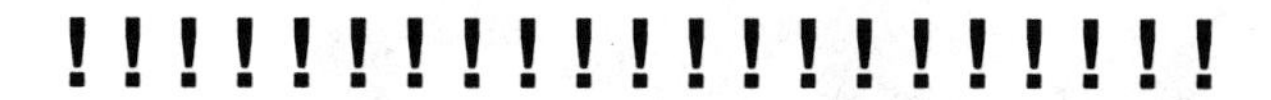

SECTION 3

SOLUTIONS

cryptic clues

```
  4 5 4 8 3 4
  4 8 3 0 6 4
        4 8 3
      4 8 8 6
  -----------
  9 4 3 2 6 7
```

It is clear from the problem that Anne is older than Amanda and that Amanda is older than Andrea. Suppose that Anne was born in the year $1900 + x$, Amanda in the year $1900 + y$, and Andrea in the year $1900 + z$, where $x < y < z$.

1. In 1989, Andrea's age was $89 - z$. When Anne was twice this age, Anne's age was $2(89 - z)$. Since Amanda is $y - x$ years younger than Anne, then at this time, Amanda's age was $2(89 - z) - (y - x)$. Since Andrea's age in 1990 was $90 - z$, we get $90 - z = 2(89 - z) - (y - x)$, or $\underline{x - y - z = -88}$.
2. When Andrea was born, Amanda's age was $z - y$. When Andrea was five years younger, Andrea's age was $z - y - 5$. Since Amanda is $z - y$ years older than Andrea, Amanda's age was $(z - y) + (z - y - 5)$ or $2(z - y) - 5$ when Andrea's age was $z - y - 5$. Since Andrea's age in 1991 was $91 - z$, it follows that $91 - z = 2(z - y) - 5$, or $\underline{2y - 3z = -96}$.
3. When Amanda was born, Anne's age was $y - x$. When Amanda was five years younger, Amanda's age was $y - x - 5$. Since Anne is $y - x$ years older than Amanda, Anne's age was $(y - x) + (y - x - 5)$ or $2(y - x) - 5$ when Amanda's age was $y - x - 5$. Since Amanda's age in 1992 was $92 - y$, it follows that $92 - y = 2(y - x) - 5$, or $\underline{2x - 3y = -97}$.

Solving this system of equations, we get $x = 19$, $y = 45$, and $z = 62$. So Anne was born in 1919, Amanda in 1945, and Andrea in 1962.

an inviting problem

```
      3 4 5
      7 0 3
    9 4 7 8
        2 4
        7 3
  ---------
  1 0 6 2 3
```

(5 and 8 interchangeable)

The "right thing" to bring to the party is anything that has a double letter in the spelling of its name.

splitting hares

```
            5 8
              4
      6 8 0 5 6
            6 8
        7 0 5 6
      3 4 1 6 9
    9 0 3 4 6 5
            8 7
      2 0 4 1 5
            8 7
  -------------
  1 0 3 3 4 6 5
```

i.d.t. – fiftyone

```
  2 9 7 1 6 2 2 8
  2 9 7 1 6 2 2 8
          6 1 3 2 2
```

```
  2 9 7 1 6 2 2 8
  2 9 7 1 6 2 2 8
        6 1 3 2 2
        6 1 3 2 2
          8 9 8 2
  ---------------
  5 9 5 6 4 0 8 2
```

i.d.t. – fiftytwo

```
    5 0 7 6 4 4 8
    5 0 7 6 4 4 8
          8 0 8 4
          8 0 8 4
            2 8 4
            2 8 4
  ---------------
  1 0 1 6 9 6 3 2
```

i.d.t. – ninetyone

5 1 0 7 5 9 2 6 3
6 0 6 3 5 3 3 6
3 0 8 1 5 3 3 6
4 0 4 5 3 3 6
5 1 7 3 3
5 1 7 3 3
2 6 3
2 6 3
6 0 6 3 5 9 2 6 3

i.d.t. – sixty

1 4 3 7 6
1 4 3 7 6
1 4 3 7 6
1 4 3 7 6
1 4 3 7 6
1 4 3 7 6
5 4 5 1
8 5 1
8 5 1
8 5 1
9 4 2 6 0

i.d.t. – fiftysix

4 0 4 3 9 3 3 4
4 0 4 3 9 3 3 4
8 0 2 3
8 0 2 3
7 0 6
1 4 3
1 4 3
8 0 8 9 5 7 0 6

prestidigitation

1 6 4 4 0
2
8 6 4 9 3 7
8 1 2 4 8 0
2 9 8 5
5 9 3
1 6 9 7 4 3 7

Answers to the questions:

1. FIFTY ONE
2. TWO HUNDRED AND FIFTY ONE
3. ONE HUNDRED AND THIRTEEN
4. FIFTY + SIXTY = ONE HUNDRED AND TEN (There may be others.)
5. 248 → 284 → 285 → 267 → 313 → (248)

wholes with holes

1 8 3 0
7 3 9 8
1 3 9 8
1 3 8 0
3 9 8
1 3 9 8
1 3 8 0 2

(a)

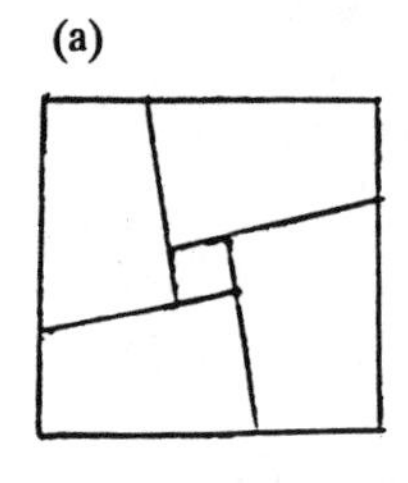

(b)

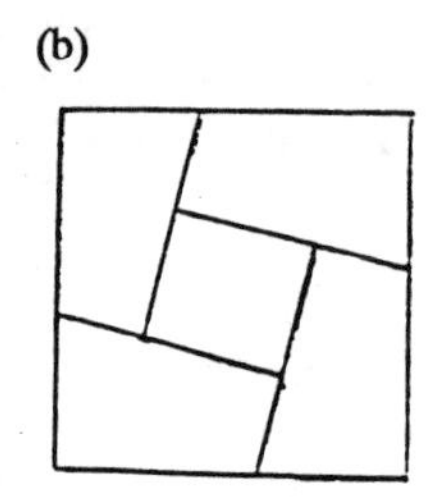

(c)

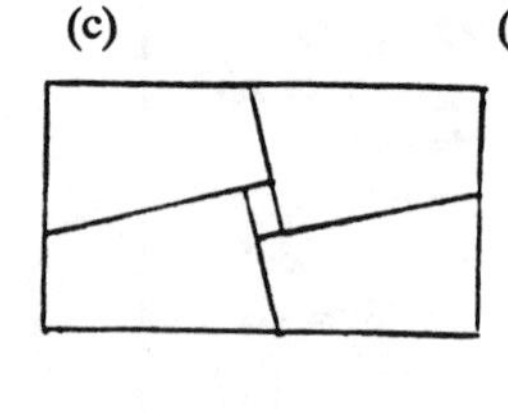

(d)

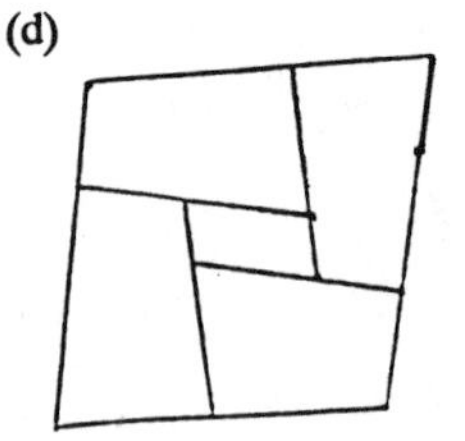

i.d.t. – thirty

```
  8 7 6 7 4
  1 2 3 7 7
    5 0 6 7
    5 0 6 7
    5 0 6 7
    5 0 6 7
-----------
1 2 0 3 1 9
```

division decisions

```
      1 4 5 4
7 4 0 6 7 4 7
  2 4 3 9 8 4
          1 4
      4 5 4 8
-------------
7 6 5 6 7 4 7
```

i.d.t. – eighty

```
1 4 8 3 5 8
1 4 8 3 5 8
1 4 8 3 5 8
1 4 8 3 5 8
1 4 8 3 5 8
1 4 8 3 5 8
      1 4 2
      1 4 2
      1 4 2
      1 4 2
-----------
8 9 0 7 1 6
```

up in smoke

```
      3 7
    5 6 4
      3 7
  9 3 8 4
    1 0 2
    1 0 0
---------
1 0 2 2 4
```

(6 and 8 interchangeable)

Harry The Hobo will have enough tobacco on hand for eleven consecutive days, as the ensuing argument shows:

The 100 cigar butts form 25 cigars. After smoking these, Harry has 25 new butts, which can be used to form an additional 6 cigars, with 1 butt left over. When these are smoked, the 6 new butts combine with the leftover one, giving Harry 7 butts with which to work. He forms 1 cigar more, with 3 butts remaining; when he smokes this cigar, Harry gets another butt. This last set of 4 butts are sufficient to create still 1 cigar more, thus accounting for 33 cigars formed in all. The result then follows by dividing this total by 3.

even odds

```
          9
  8 4 8 5 2
  8 4 8 5 2
-----------
1 6 9 7 1 3
```

The paragraph contains twelve heteronyms:

THE OBJECT OF THIS QUIZ IS TO RECORD ALL OF THE HETERONYMS THAT APPEAR IN THIS PARAGRAPH. WHERE DOES THIS SEARCH LEAD? WELL, IF YOU READ VERY, VERY CAREFULLY, AND YOU USE ALL OF YOUR POWERS OF OBSERVATION, YOU SHOULD HAVE NO EXCUSE BUT TO FIND THE TOTAL NUMBER OF HETERONYMS PRESENT HERE. IF YOU WIND UP GETTING THEM ALL, TAKE A BOW; IF NOT, DON'T SHED A TEAR!

i.d.t. – twenty

```
505153
 69755
 69755
   235
   235
   235
------
645368
```

i.d.t. – fifty

```
10207
10207
36800
36800
  307
  307
  307
-----
94935
```

(2 and 8 interchangeable)

optical allusion

```
230829
230829
------
461658
```

A 90° counterclockwise rotation turns the frog into a horse.

sesquipedalia

```
    54
   341
 89757
 14625
    84
   897
------
105758
```

(3 and 6 interchangeable)

The proverbs:

1. The early bird catches the worm.
2. All that glitters is not gold.
3. A rolling stone gathers no moss.
4. Haste makes waste.
5. A fool and his money are soon parted.
6. Look before you leap.
7. People who live in glass houses shouldn't throw stones.
8. Too many cooks spoil the broth.

it speaks for itself

```
   724
   724
  6230
 96202
   302
 96022
------
200204
```

The three spelling mistakes:

1. The word "HERE" is misspelled H–E–A–R.
2. The word "ERRORS" is misspelled E–R–R–E–R–S.
3. The work "TWO" is misspelled T–H–R–E–E.

i.d.t. – thousand

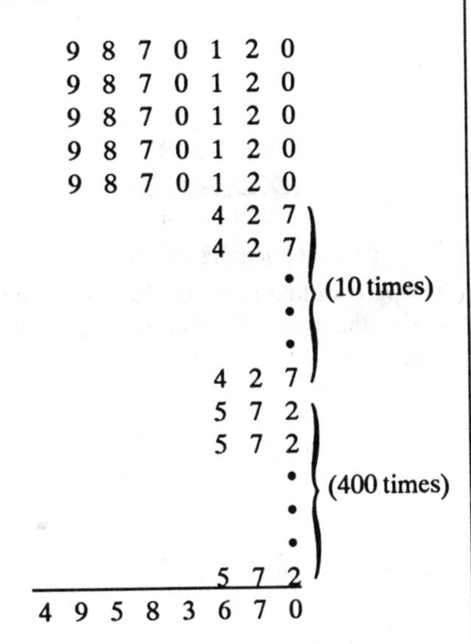

go for the gold

				2	4
	1	9	2	5	9
				2	7
			6	9	0
	8	2	3	9	6
1	0	2	3	9	6

(4 and 7 interchangeable)

To find the bag that contains the real gold nuggets, assign the digits 1 through 9 to the nine bags. Then remove nine nuggets from bag #1, eight nuggets from bag #2, seven from bag #3, . . . , and finally one from bag #9. When the forty-five withdrawn nuggets are placed on the pan of the scale, their total weight will be 44.n ounces. Hence, bag #n is the one that contains the real gold.

i.d.t. – eightyseven

2 4 8 0 1 5 + 6 2 9 2 3 = 29(1 0 7 2 2)

the vanishing square

	2	0	8	9	0
			3	2	7
	3	7	7	3	0
	6	8	9	5	1
			4	8	4
					8
1	2	8	3	9	0

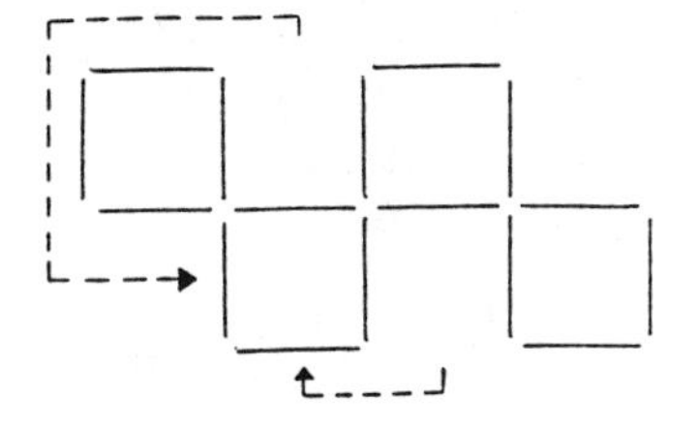

cover

						1	9
				2	1	6	9
	8	4	5	3	0	8	0
		9	3	9	1	2	6
		6	8	5	3	4	0
1	0	0	7	9	7	3	4

i.d.t. – sixtysix

1	8	3	0	1	6	6	4
1	8	3	0	1	6	6	4
1	8	3	0	1	6	6	4
1	8	3	0	1	6	6	4
				4	3	4	6
					9	4	6
					9	4	6
					9	4	6
					9	4	6
					9	4	6
7	3	2	1	5	7	3	2

i.d.t. – fortytwo

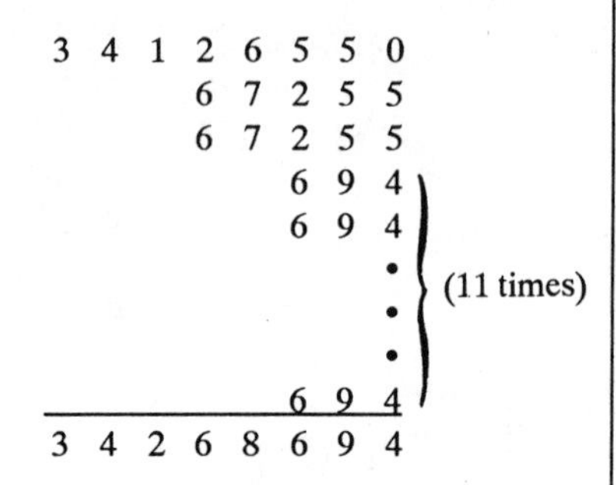

everything's relative

```
      4 0 8
      1 6 9
          2
  8 5 2 3 3
1 2 5 4 3 7
-----------
2 1 1 2 4 9
```

(0 and 6 interchangeable)

The party of four consists of a brother and sister, the brother's daughter, and the sister's son.

i.d.t. – seventy

```
  9 5 6 7 9 4
  6 1 6 0 6 7
  6 1 6 0 6 7
  6 1 6 0 6 7
  6 1 6 0 6 7
    9 2 8 6 6
    9 2 8 6 6
-------------
3 6 0 6 7 9 4
```

i.d.t. – fortyone

```
9 7 3 8 0 4 4 5
    4 1 4 2 4 5
          7 5 4 )
          7 5 4 |
              . | (16 times)
              . |
              . |
          7 5 4 )
---------------
9 7 8 0 6 7 5 4
```

walking the plank

```
    8 2 6 7
          6
2 6 0 0 5 1
        7 4
  9 1 6 3 2
      7 2 1
-----------
3 6 0 7 5 1
```

Tell Pirate Pete to place one of the boards so that it becomes the hypotenuse of an isosceles right triangle whose legs are adjacent sides of the shoreline. If he were then to carry the second board and walk (carefully!) to the midpoint of the first, the second board would be sufficiently long to reach the corner of the island.

i.d.t. – ninety

```
  8 7 0 5 9
  2 7 3 4 5
  5 4 6 2 2
    1 7 1 2
      5 2 1
-----------
1 7 1 2 5 9
```

(0, 3, and 6 interchangeable)

dedication

```
          8 6 0
    5 1 6 8 2 6
5 6 2 2 3 9 3 4
---------------
5 6 7 4 1 6 2 0
```

for the cruciverbalists

```
      3 4 5
      4 2 0
      4 7 7
      4 7 7
  9 5 4 6 6
    8 4 5 7
-----------
1 0 5 6 4 2
```

solution to the crossword:

C	C	C	C
U	U	U	U
B	B	B	B
E	E	E	E

open and shut case

```
    9 0 8 5
  1 9 9 6 7
4 2 9 7 8 1
4 2 9 7 8 1
  1 9 9 6 7
-----------
9 0 8 5 8 1
```

The number of times that the status of a door is altered is equal to the number of distinct divisors of the locker number. The number of such divisors will always be even, except when the locker number happens to be a perfect square. Since each door is closed at the outset, an even number of alterations will keep it closed, while an odd number of alterations will leave it open. Therefore, the problem reduces to finding all perfect squares that are less than or equal to 1000. Since $(32)^2$ is the smallest perfect square that exceeds 1000, there are exactly 31 doors ajar at the end of the procession.

i.d.t. – thirtythree

2(1 6 1 0 1 8) + 2 1 8 + 7 8 1 =
2 9 3 5 2 4 + 2 9 5 1 1

i.d.t. – hundred

```
    5 1 6 9 9 \
    5 1 6 9 9  |
            •  |  (20 times)
            •  |
            •  |
    5 1 6 9 9 /
      4 8 3 9 \
      4 8 3 9  |
            •  |  (8 times)
            •  |
            •  |
      4 8 3 9 /
-------------
1 0 7 2 6 9 2
```

count on it!

```
    6 4 1 2
      7 1 4
    5 3 8 6
        3 8
  9 1 3 6 2
-----------
1 0 3 9 1 2
```

The unique sequence of length ten is 6 2 1 0 0 0 1 0 0 0.

i.d.t. – sixtyone

```
2 0 2 3 6 3 3 2
6 1 0 5 6 3 3 2
  9 0 4 6 3 3 2
      6 1 5 3 3
      6 1 5 3 3
          8 2 3
          8 2 3
          8 2 3
          8 2 3
          8 2 3
          8 2 3
          8 2 3
---------------
9 0 4 6 7 8 2 3
```

high-powered equations

```
      4 9 3
    1 5 5 8
  6 7 5 5 1
  6 7 5 5 1
        5 1
-----------
1 3 7 2 0 4
```

i.d.t. – forty

```
1 5 4 5 7
1 5 4 5 7
1 5 4 5 7
1 5 4 5 7
1 5 4 5 7
  8 9 4 5
---------
8 6 2 3 0
```

urban affairs

```
    8 9 1 2
    8 7 2 1
        6 1
  8 9 6 2 1
-----------
1 0 7 3 1 5
```

Fifty -city words appear below. No doubt others exist.

1. analyticity
2. atrocity
3. audacity
4. authenticity
5. capacity
6. catholicity
7. causticity
8. centricity
9. complicity
10. concentricity
11. contumacity
12. domesticity
13. duplicity
14. eccentricity
15. edacity
16. egocentricity
17. elasticity
18. electricity
19. ethnicity
20. ethnocentricity
21. felicity
22. ferocity
23. incapacity
24. inelasticity
25. intercity
26. loquacity
27. mendacity
28. multiplicity
29. opacity
30. overcapacity
31. paucity
32. periodicity
33. perspicacity
34. pertinacity
35. plasticity
36. precocity
37. publicity
38. pugnacity
39. rapacity
40. reciprocity
41. sagacity
42. scarcity
43. simplicity
44. specificity
45. tenacity
46. toxicity
47. velocity
48. veracity
49. vivacity
50. voracity

solutions chart

The following symbols are used in the chart:

- * each solution has an interchangeable pair of digits
- ** each solution has two interchangeable pairs of digits
- *** each solution has an interchangeable trio of digits

alphametic	number of solutions
cover	one
dedication	one
narrative	
even odds	three
go for the gold	one*
wholes with holes	one
division decisions	two
optical allusion	one
open and shut case	two
urban affairs	one
splitting hares	one
for the cruciverbalists	one
count on it!	one
the vanishing square	two*
high-powered equations	one
sesquipedalia	one*
walking the plank	eight
it speaks for itself	one
cryptic clues	one
everything's relative	one*
prestidigitation	one
up in smoke	one*
an inviting problem	three*

alphametic	number of solutions
ideal doubly-true	
i.d.t. – ninety	one***
i.d.t. – twenty	one
i.d.t. – sixtyone	one
i.d.t. – fiftyone	two
i.d.t. – thousand	one
i.d.t. – thirty	one
i.d.t. – fortytwo	two
i.d.t. – sixtysix	one
i.d.t. – fiftytwo	one
i.d.t. – seventy	one
i.d.t. – ninetyone	one
i.d.t. – hundred	one
i.d.t. – eightyseven	two**
i.d.t. – sixty	two
i.d.t. – forty	three
i.d.t. – fiftysix	two
i.d.t. – eighty	one
i.d.t. – thirtythree	four
i.d.t. – fifty	two*
i.d.t. – fortyone	one

about the author

Steven Kahan has been teaching mathematics at Queens College of the City University of New York since September 1970. For the past fifteen years, he has served as the alphametics editor of *The Journal of Recreational Mathematics*, one of the foremost publications in its field.

Throughout the years, he has created over sixteen hundred original alphametics, forty-two of which appear in his first collection of such puzzles, entitled *Have Some Sums To Solve*. He has also authored an intermediate algebra text, published in 1981 by Harcourt Brace Jovanovich, Inc.

He lives happily with his wife and two children in Hollis Hills, New York.

Journal of
RECREATIONAL MATHEMATICS

AIMS & SCOPE

Editor Joseph Madachy invites you to take a look at the lighter side of mathematics.

The *Journal of Recreational Mathematics* is thought-provoking and stimulating — packed with geometrical phenomena, alphametics, solitaires and games, chess and checker brainteasers, problems and conjectures, and solutions.

The *Journal of Recreational Mathematics* offers everyone interested in math a never-ending parade of the exciting side of numbers.

Subscription Information: ISSN 0022-412X
Price per volume — 4 issues yearly
Institutional Rate: $78.00
Individual Rate: $18.95
Postage & handling: $4.50 U.S. & Canada; $9.35 elsewhere.

Complimentary sample issue available upon request

RM11/92

Baywood Publishing Company, Inc.
26 Austin Avenue, P.O. 337, Amityville, NY 11701
Phone (516) 691-1270 Fax (516) 691-1770 Orders only — call **toll-free** (800) 638-7819